Honeybee Anatomy Brought to Life
Graham Kingham

ISBN: 978-1-914934-12-4

Published by Northern Bee Books © 2022

Northern Bee Books, Scout Bottom Farm
Mytholmroyd, Hebden Bridge, HX7 5JS (UK)
www.northernbeebooks.co.uk
Tel: 01422 882751
Book design by www.SiPat.co.uk

Honeybee Anatomy Brought to Life
Graham Kingham

Dedicated in memory of my parents, Donald and Joyce
Kingham and my in-laws, Frank and Doreen Clarke

Contents

Introduction 9

Egg, Larva and Pupa

    Egg 12
    Larva 15
    Head and thorax 18
    Moulting 21

The endocrine glands 23

Brain 24

Nerves 25

Oenocytes, excretory cells and fat bodies 26

Digestion 28

Malpighian tubules 31

Tracheal system 33

Muscles 35

Heart and blood 36

Drone production and development 39

Silk wrappings make strong bonds 44

Pupal stage 47

Adult bees

Respiration 52
   First, shave your bee 53
   Drawing breath 56

The Ventral Nervous System 59
   The Control centres, the Ganglia 60

Circulation 65
   Heart of the matter 66

Exoskeleton 70
   How the bee got its hump 71
   An hourglass figure 73
   Hairy beasts 75
   Armour and anchors 77

Excretory system 81
   Who was Malpighi? 82
   Expansion vessels 85

Digestion 87
   Mandibles are made for many things 88
   Proboscis, a drinking straw by another name 90
   Proventriculus: before the belly 94
   Revenge 96

Reproduction 98
   Honeybee drones, possessors of some very specialised equipment .. 99
   The life of the sperm 102
   Nature's production lines 106

Flight 109
   Diaphanous wings 110
   Morphometry or Morphology? 113

Muscles 119
   The powerhouse of flight 119

Locomotion 122
   Designed for all events 123
   The bee's knees 127

Endocrine system 129
    Fat bodies explained 130
    Glands 132
    Waxing Lyrical 138
    Martyr to the cause 141

The Sensory Organs 143
    Antennae, horizontal masts 144
    Organs of Jacobson and Johnson, smell and the hearing bee 149
    Photon collectors 151
    An opportunity to see in ultraviolet! 154
    Little eyes 158

Pests 160
    Uninvited guests 161
    Let us hear it for the Wax moth 164
    The common wasp, *Vespula vulgaris* 166

Glossary 167
References 175
Acknowledgements 179
Technical 180
Index 182
Other books by Graham Kingham 184

Fig 1. Are you sitting comfortably?

Introduction

This book started as a series of anatomy articles in the Devon Beekeeping magazine, as my interest in the microscope and anatomy grew after taking the British Beekeepers Association (BBKA) 'Microscopy' exam. The role of a microscopist in beekeeping is to help and advise their fellow beekeepers.

Anatomy is not a popular subject and is no longer taught, except for the BBKA exam system, but fascinating none the less. The purpose of this book is to show, micrographs alongside the drawings, the external and internal anatomy and the visual aspects, which do not always seem like the drawings when dissecting the bee.

By using modern photographic and software techniques, I have been able to show in detail the fascinating images.

The book is intended to be an aid for those who are interested in insect anatomy, to be used alongside the great textbooks of the masters who have stood the test of time, and who go into detail and explanations, such as Dade, Goodman, Nelson, Snodgrass and Stell, these works have inspired me.

The book is divided into three sections – larval and pupal stage of development; adult bees and pests affecting the honeybees. Some of the anatomical parts, such as the heart, have the same properties in each stage of development, this has been discussed in each section to help the reader understand without referring back, and to enable them to read and refer to each section as a separate article.

Egg, Larva and Pupa

Egg (noun.)

"The body formed in the females of all animals (except for a few of the lowest type) in which by impregnation the development of the foetus takes place". mid-14c., *egge*, mostly in northern England dialect, from Old Norse *egg*.

Larva (noun.)

1650, "A ghost, spectre, disembodied spirit" (earlier as *larve*, c. 1600), from Latin *larva* (plural *larvae*), earlier *larua* "ghost, evil spirit, demon," also "mask," a word from Roman mythology, of unknown origin.

Crowded out in its original sense by the zoological use (1768) which began with Linnaeus, who applied the word to immature forms of animals that do not resemble, and thus "mask," the adult forms.

Pupa (noun.)

"Post-larval stage of an insect," 1773, special use by Linnæus (1758) of Latin *pupa* "girl, doll, puppet" on the notion of "undeveloped creature." Related: *Pupal*; *pupiform*.

Egg

The queen lays the egg into the bottom of the cell and it is fastened upright; the head end is laid uppermost. When the egg is fully developed it passes a duct inside the queen's body from which a few sperm passes, at the top of the egg there is an opening called the micropyle through which the sperm then enters to fertilise the egg nucleus that sits near the top. The egg is about 0.33mm in diameter and 1.5 mm long, tapering down to the base, both ends are rounded; they are slightly curved inwards and weigh around 0.13 mg.

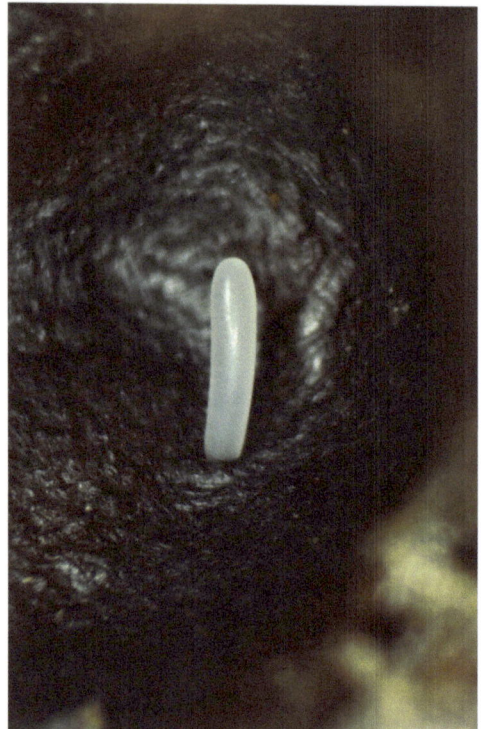

Fig 2. Egg in the bottom of an old cell, note the darker wax bottom. X 40 Magnification.

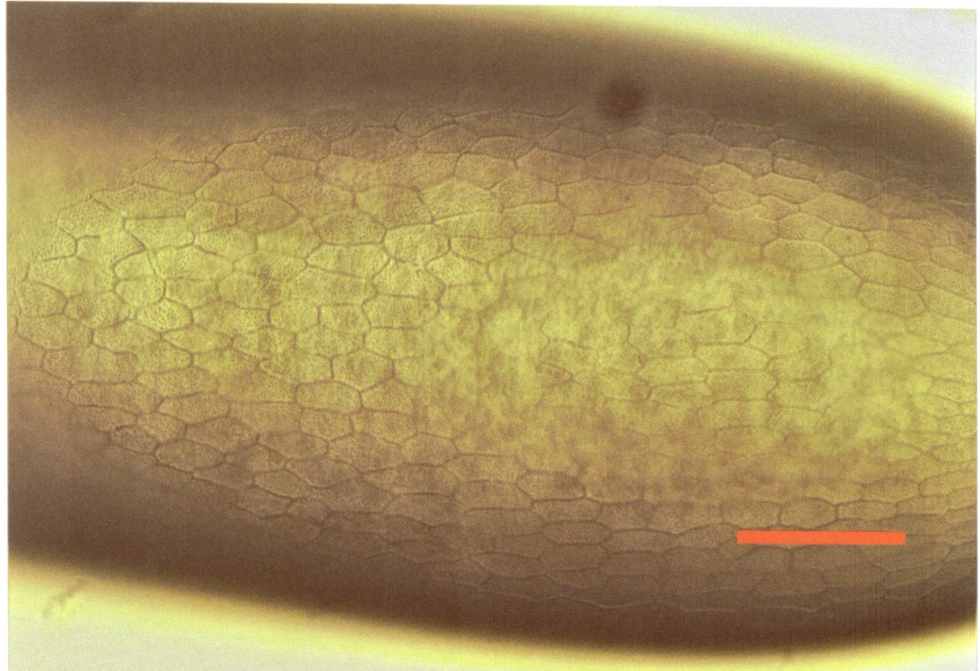

Fig 3. Reticulated chorion surface of a day old egg, meaning 'marked like a net'. Scale bar 100 microns. X 100 Magnification.

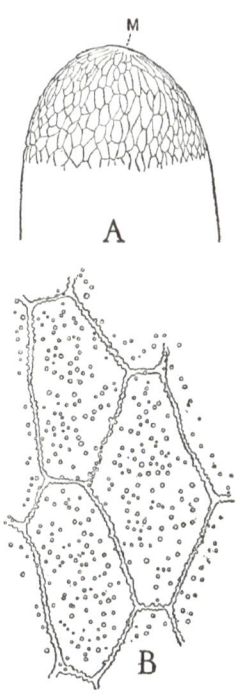

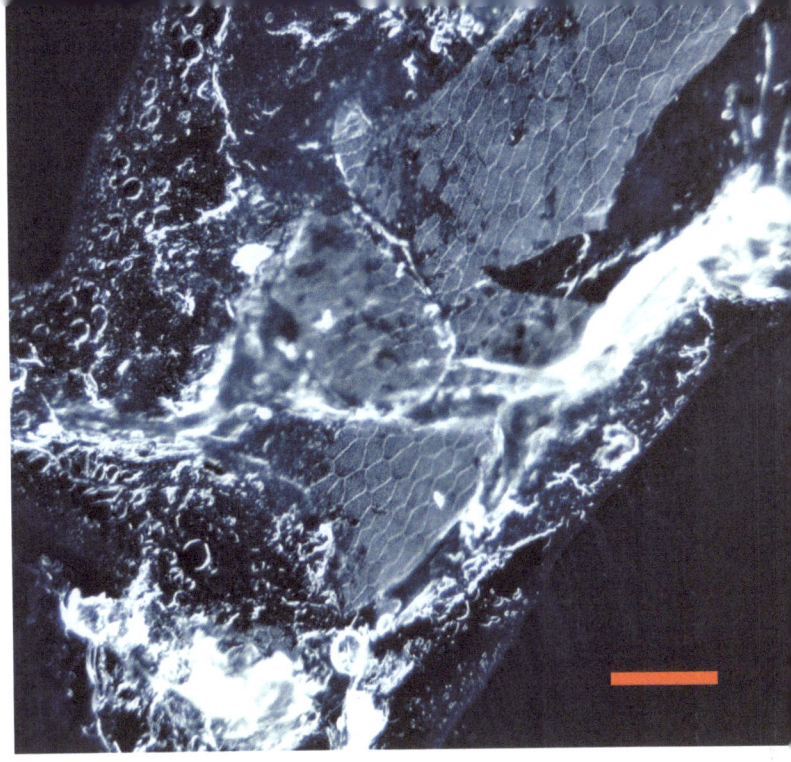

Fig 4 (above). [A] Anterior surface view of egg, showing reticulated chorion and micropylar area. X 100 magnification. [B] Magnified X 713 chorion showing tubules. Taken from Nelson.

Fig 5 (above right). The surface of an egg showing the hexagonal pattern that was imprinted from the basic cell from where it was formed. Scale bar 100 microns. X 100 Magnification.

Fig 6 (right). Stages in development from egg to pupa. Taken from Dade, by kind permission of IBRA.

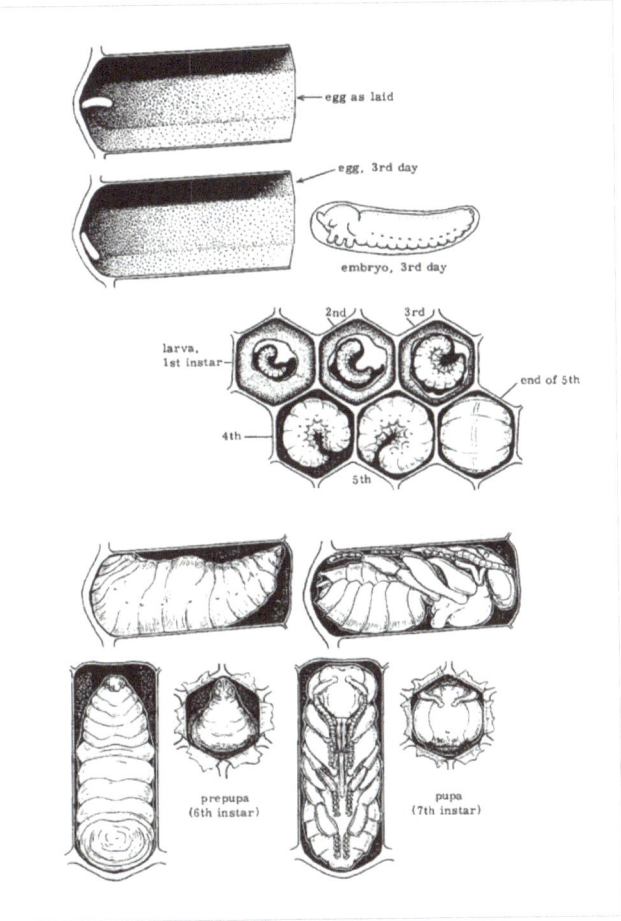

The outer layer is made from the chorion, a membrane-like shell and has a hexagonal pattern imprinted on its surface, caused by the surrounding tissue where it was formed inside the queen's reproductive tract. This outer layer protects the egg for three days whilst it hatches. The inner membrane is made from vitelline, the true cell wall; this has a layer of cytoplasm next to it, holding the yolk contents and nucleus.

J A Nelson in his book 'The Embryology of the Honeybee' describes the development of the head and thorax in the embryo in three stages. The first reflects the worm-like ancestor at 42 to 44 hours development, showing paired appendages on the front segments. Stage two - at 58 to 60 hours the mouth and labrum move toward the top of the embryo and the leg segments join. Stage three - the labium fuse together, the Stomodaeum, foregut, and the mouth then form. The larva will hatch at the end of day three after being laid. At this stage, the attending nurse bee will bathe it in royal jelly.

Fig 7. Newly laid egg. Green arrow, nucleus. Yellow arrow, yolk. Blue arrow, Chorion. Black arrow, Vitelline membrane. Red arrow, Cytoplasm. Purple arrow, Micropyle.

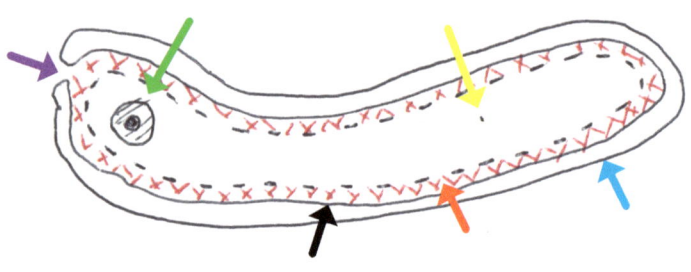

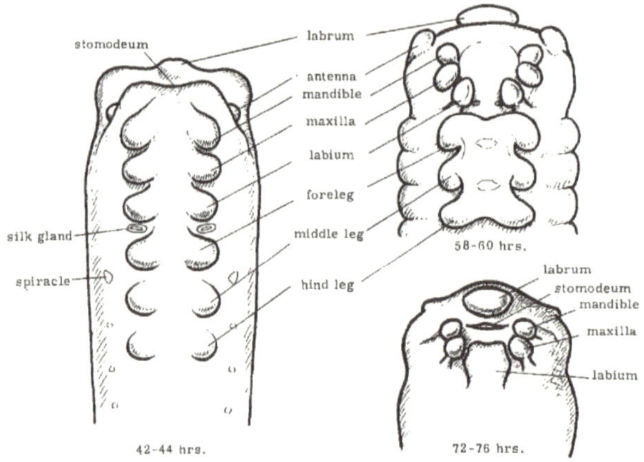

Fig 8. The development in a time frame of the embryonic mouthparts, showing the final fusion of the two labia. Taken from Dade, by kind permission of IBRA.

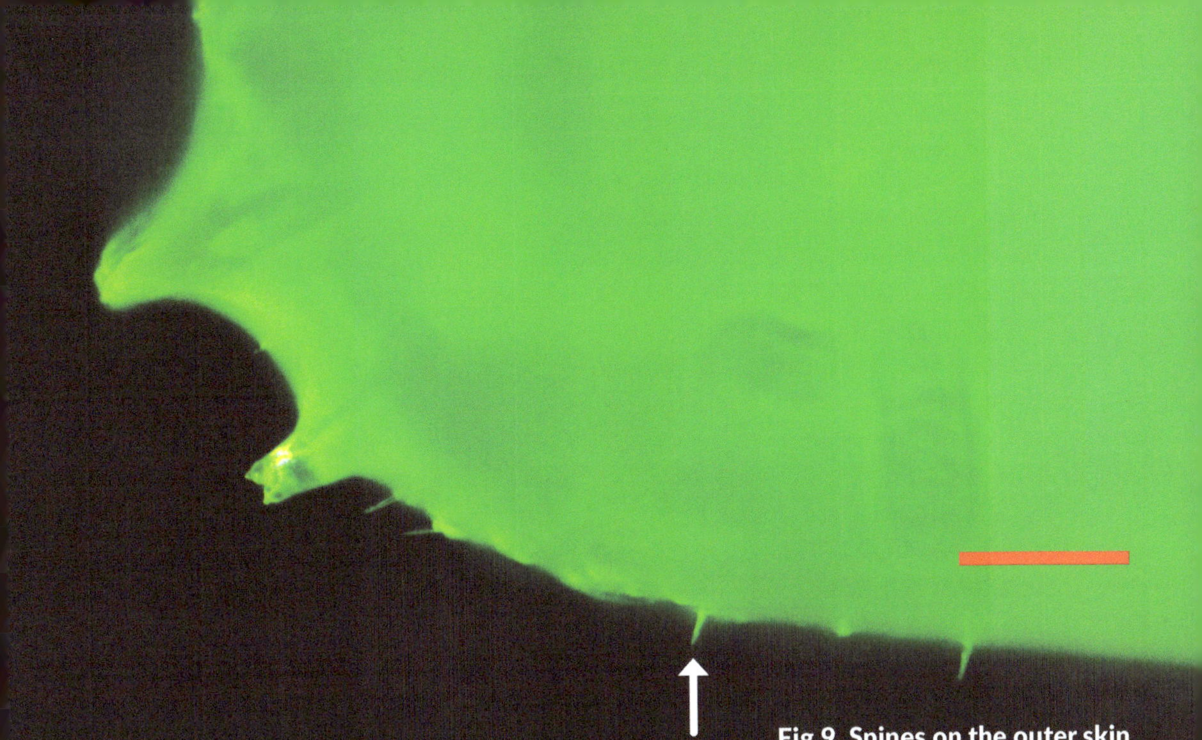

Fig 9. Spines on the outer skin layer of a larva. Scale bar 100 microns. 100 X Magnification.

Larva

The word Larva means 'disguise or ghost'.

Larvae of honeybees, unlike most other insects, are not born with legs or eyes to move about or to feed themselves. They possess a mouth that is only able to accept semi-liquid food and so spend their development life being fed by young worker bees.

The larval body is divided into segments except for the head. The internal organs also show their segmental origins.

The bee larva has a fusiform shape, taken from the Greek, meaning 'a spindle that has tapering ends'. The posterior end is smaller. Its colour is ivory white. The larva is divided by constrictions into a head and thirteen segments, three of which belong to the thorax, the remainder to the abdomen. The ventrolateral region of abdominal segments one to nine are raised to form a series of rounded swellings or side projections called the epipleural lobe.

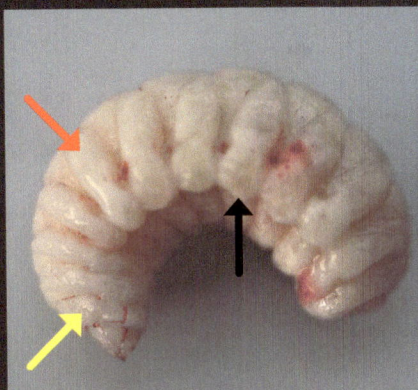

Fig 10. Stained and fixed larva. Yellow arrow start of thorax, red arrow start of the abdomen, black arrow epipleural lobe.

The term "hatching spine" is used for a variety of cuticular structures found on the embryonic cuticle of the first instars of the honeybee. They are situated just above the spiracles near the head end. These are used to cut through the embryonic cuticle or chorion at the time of hatching from the egg.

Fig 11. Larval stages.

A single larva will eat about 25 milligrams of royal jelly during its larval stages. With up to 200,000 eggs laid in a colony in a bee season, this equates to about five litres per year.

There are three types of food fed to larvae and a dedicated mix is given to each caste, which affects their development in different ways. A white food secreted by the mandible glands, also known as royal jelly, is fed mainly to queens (Remember the queen comes from an ordinary fertilised egg that would normally develop into a worker bee. Royal jelly helps develop the ovaries and switches off the development of other body parts such as the mandibles, food glands, wax glands, scent glands and pollen carrying parts on the hind legs). Secondly, a clear food from the hypopharyngeal glands is fed to both queen and worker larvae, and thirdly a yellow food derived from pollen is a form of protein and used for growth.

Day one to three the queen is fed white brood food, then the last few days fed clear and white at a ratio of 1:1

Worker larvae are fed a mixture of white, clear and yellow ratio 2:9:3

Larval developmental times. The stadia of a worker bee.

Approximate growth stages	Instar (Stages of growth)	Ecdyses (Moulting stages)
Larva	1st ½-3/4 day	1st
Larva	2nd	2nd
Larva	3rd	3rd
Larva	4th	4th
Larva/Prepupa	5th	5th
Pupa	6th	6th
Imago	7th	7th

The larva lies on its side on the floor of the cell in a pool of brood food, which they can ingest at their leisure and later in their development; raw pollen is fed to them providing protein, all supplied by the nurse bees.

The spiracles are found on each side, but only those on the top side are exposed to the air. Those on the bottom are in the larval food. Hence, when transferring worker larvae for queen rearing it is important not to flip the larvae, as they will not be able to breathe.

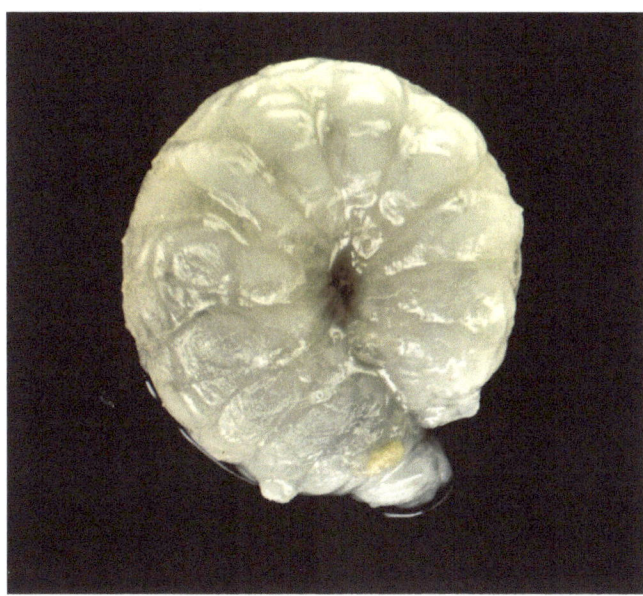

Fig 12 (above). Mature Queen pupa is large over 12 mm long. The multiple queen cells that workers build during swarming and supersedence have a dedicated position on the outside of the comb, pointing downwards, unlike the horizontal cell of the workers and drones. Fig 13 (left). Larva removed from the pool of brood food, between five and six days old.

The head and thorax

The head is conical in shape and joined to the first thoracic segment. It contains the brain, ganglia and hormone-producing glands. The facial features of the larva continue to develop into the full-grown adult bee.

All of the mouthparts are soft and fleshy and covered only by thin chitin, except for the small spines described below.

Beneath the labrum is the mouth opening on each side of which are the conical mandibles curved in such a way that their pointed tips lie beneath the labrum.

The maxilla, the lower jaw, which is attached to the rear part of the head capsule, forms a chin like structure and continues forward to wrap around the upper part of the labium where it forms a large diameter rounded tube on each side, tipped with a small spine.

The labium, meaning 'lower lip', is relatively large, projecting beyond the other mouthparts. It bears the slit-like common opening of the silk glands, which will later be used to spin the cocoon.

The upper surface of the labium is separated from the lower surface of the labrum by a narrow cleft, forming the mouth opening, which is bounded laterally by the mandibles and maxilla.

A pair of mandibles (upper jaw) sits on top of each of the maxilla, mimicking their shape but is smaller and also tipped with spines.

Above these structures is the nose-like feature, the labrum, or upper lip.

The tips of the antennal rudiments project slightly above the general surface of the head showing small rounded buds.

The spines have been given histological and electron microscopical examinations, which have shown that they are composed of coupled, paired sensilla, connected via nerves to the suboesophageal ganglia. Further investigation has shown that the sensilla contain sugar and salt receptors, that help identify food sources such as sugars in the form of brood food and pollen that is rich in protein and minerals.

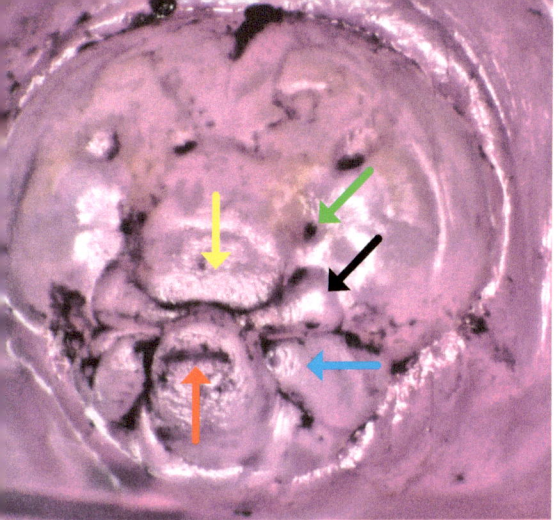

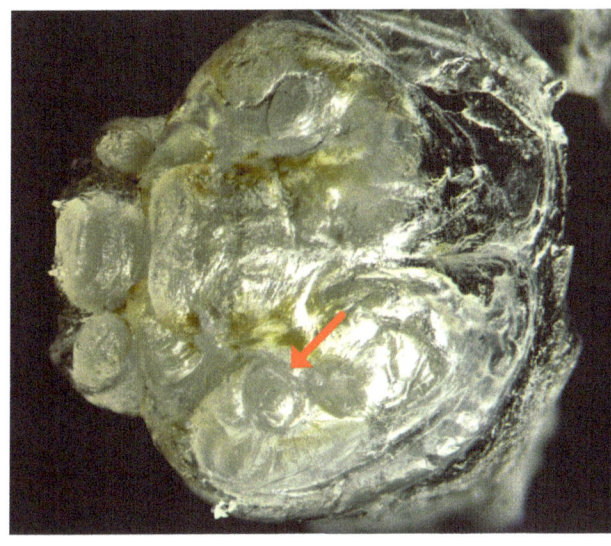

Fig 14. Stained larval head. Labium, red arrow showing silk glands opening. Maxilla, blue arrow. Labrum, yellow arrow. Mandibles, black arrow. Tentorium pit, green arrow. X 40 Magnification.

Fig 15. View of the top of the head, showing antenna bud.

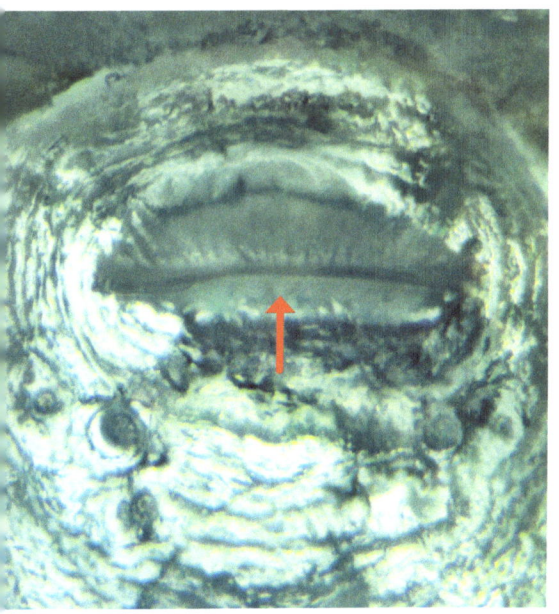

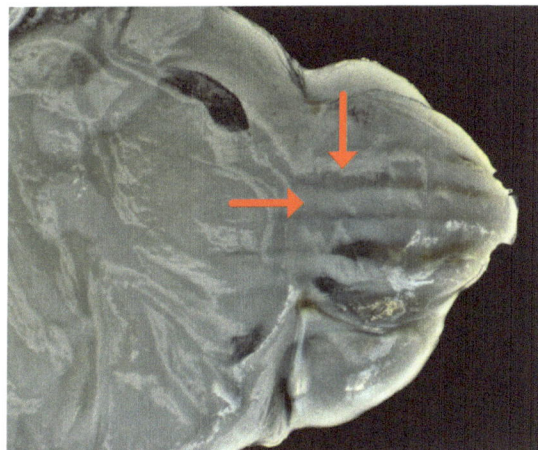

Fig 17. Head view of developing larva showing mouthpart development.

Fig 16. Enlarged view of silk gland opening.

Body tegument

The body wall consists of a single epithelial layer of small cells, the epidermis, clothed externally by a delicate cuticle. The epidermis differs greatly in thickness in different parts of the body but its average thickness is greatest in the head. The cuticle is also thicker here and more rigid than elsewhere.

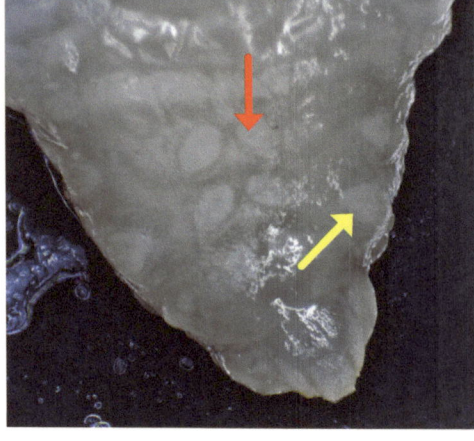

Fig 18. Legs development area, red arrow; wings development area, yellow arrow.

The antennal rudiments are ovoid and situated in deep peripodal cavities.

The leg rudiments are ovoid, like the antenna, and are situated in deep open depressions covered externally only by the cuticle. The wing rudiments are flat hollow outgrowths of the hypodermis of the mesothoracic segments and are situated in shallow depressions low down on these segments; the wing rudiments are heart-shaped.

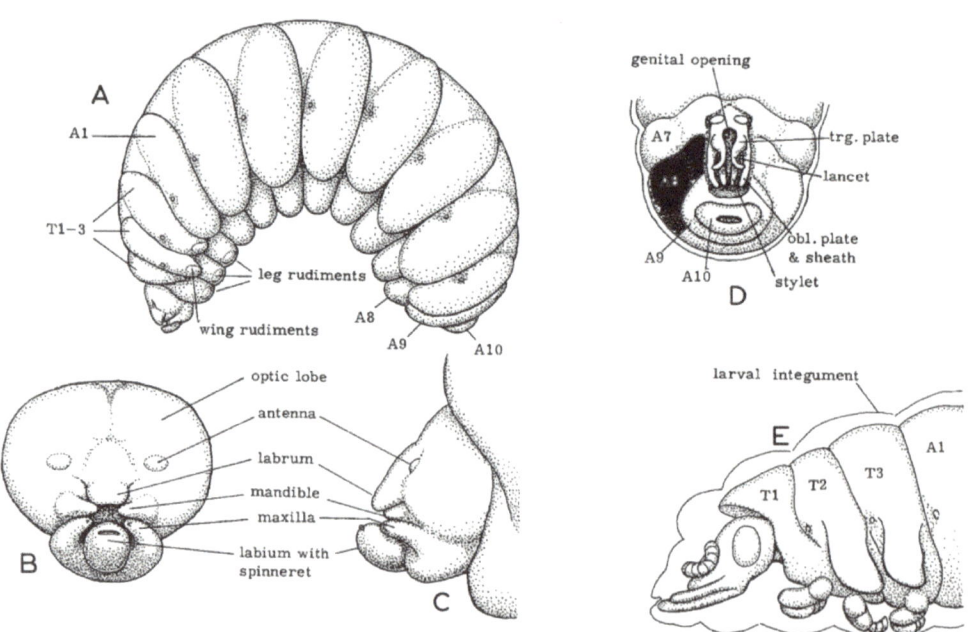

Fig 19. External anatomy of larva and prepupa. A. Larva. B. Face of the larva. C. Head left side. D. Posterior segment of prepupa, sting detail. E. Head and thorax of prepupa. Showing reorganisation of head and thorax during development. Taken from Dade by kind permission of 'International Bee Research Association' (IBRA).

Moulting

Growth in the larva is very rapid during the five developmental stages of the larvae; they eat a prodigious amount of food for five days increasing their body weight by a factor of 1000. This translates in human terms as a newborn baby at five days old, weighing 1.5 tons.

Because their outer cuticle especially that of the head capsule does not grow during this rapid stage of growth, the cuticle has to be shed to accommodate the increasing mass. The stages of growth are called Stadia, meaning 'state or position', and Ecdysis (Greek for to strip or moult) which is when the old cuticle is shed. This regulation of growth is controlled by the endocrine system.

The moulting process consists of the secretion of a moulting fluid that softens and dissolves the inner layer of the cuticle, thus freeing the old outer exocuticle; the epidermis then secretes a new cuticle layer. The action of breathing helps the larva to rupture the old cuticle and the outer layer is eventually sloughed off. After the sixth moult, the cuticle of the larva that was thin and almost transparent becomes sclerotised - hard and coloured.

The final membrane can be seen on an emerging bee.

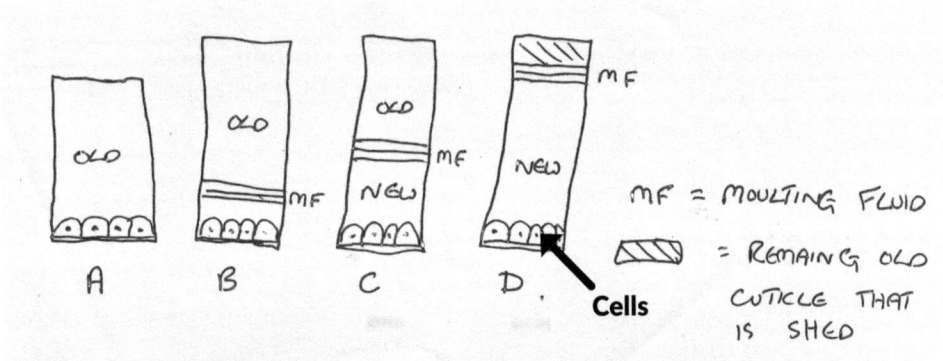

Fig 20. A. Normal cuticle. B. Start of moulting. C. New cuticle developing. D. Final stages of moulting, old skin waiting to be sloughed off.

Once feeding has ended and when the cell is being sealed the larva changes position and now sits with its head towards the sealed end of the cell, it begins to spin a web from the spinneret finishing about 24 hours later.

There are little external differences during the first four moults, after which the larva makes its most growth, adding up to 40% of its weight. About two days after this moult it receives its last feed and the cell is capped.

Fig 21. Pupal cuticle from a newly emerging worker bee.

At this stage, the connection between the ventriculus and hindgut is opened and the Malpighian tubules are opened into the hindgut, allowing excreta to enter the hindgut and to be voided into the cell. This is later mixed up with the silk threads when making the cocoon.

The different shapes of growth, which are quite visible, are called the Instar- Latin for bigness.

The moulting process occurs when enzymes are released, causing softening and dissolving of the inner layer of the cuticle that then frees the outer exocuticle, which is shed; the epidermis then secretes a new cuticle layer. The sixth moult occurs when the motionless prepupa, which is the final stage of the fifth larval Instar, moults into a pupa.

The pupa then develops into the adult, which can be seen by the tanning and sclerotisation of the adult cuticle under the thin pupal cuticle. Shortly before emerging from the brood cell, the final adult moult takes place, and in newly emerged bees, pieces of the shed pupal cuticle can still be seen clinging to the bee's body.

The endocrine glands

These are ductless glands that make hormones, Greek for 'setting in motion,' which are secreted into the body and act as biochemical controls for such processes as growth, development and bodily maintenance. The Juvenile hormone (JH) is produced and secreted by a pair of glands, the corpora allata, which sit at the rear end of the brain, and Ecdysone, which is produced by the prothoracic gland. While Ecdysone controls and synchronizes the timing of the moulting events, it is JH that determines the moulting type. For a moult from one to the next larval stage to occur the JH level in the haemolymph needs to be high, while for the pupal and adult moults it must be at basal levels.

The neurosecretory cells are found in the brain and ganglia. The corpora cardiaca, meaning bodies near the heart, are two glands that sit behind the brain connecting to it via nerves containing the axons of neurosecretory cells of the brain (pars intercerebralis).

The paired Prothoracic glands are found near the first spiracles and are responsible for secreting the hormone Ecdysone that stimulates moulting. It is under control from other hormones, namely the prothoracicotropic hormone secreted from the corpora cardiaca attached to the brain.

The other important pair of glands are the corpora allata, 'meaning body moved to their final position' during development. Each corpus allatum is about 0.85 mm in diameter. They sit above the oesophagus, right behind the brain, to which they are connected via a pair of nerves.

Once sealed the larva changes position and sits with its head at the sealed end of the cell, it then spins a web from the spinneret finishing about 24 hours later. It then stays still for 24 hours, when it goes through the prepupal Instar.

The growth rate of the larva can be affected by temperature and food supplies.

The corpora allata persist in the adult bee and are important for regulating the adult life cycle of the workers, but the prothoracic glands degenerate at the end of metamorphosis.

Brain

The central nervous system consisting of the brain and the ventral ganglia (ganglia are dependent control centers) are some of the few organs that are retained by the larva and not completely dissolved during pupation. However during the pupal stage, the ganglia are reduced down from eleven to seven, four merge in the thorax, and the sixth and seventh abdominal ganglia merge. The larval brain is smaller than that of the adults, and during the pupal stage, the optical and antennal lobes and nerves have yet to be fully developed.

Neurosecretory cells are found in the brain and ganglia. The corpora cardiaca, are two glands that sit behind the brain connecting to it via neurosecretory cells.

The brain is relatively large in size to the rest of the body. It is divided into two crescent-shaped halves, the slender end of which forms two pairs of nerve fibres called the crura cerebri, the first of which is the antennal nerve. The suboesophageal ganglion gives rise to four pairs of nerves that serve the mandibles, maxillary and labial areas.

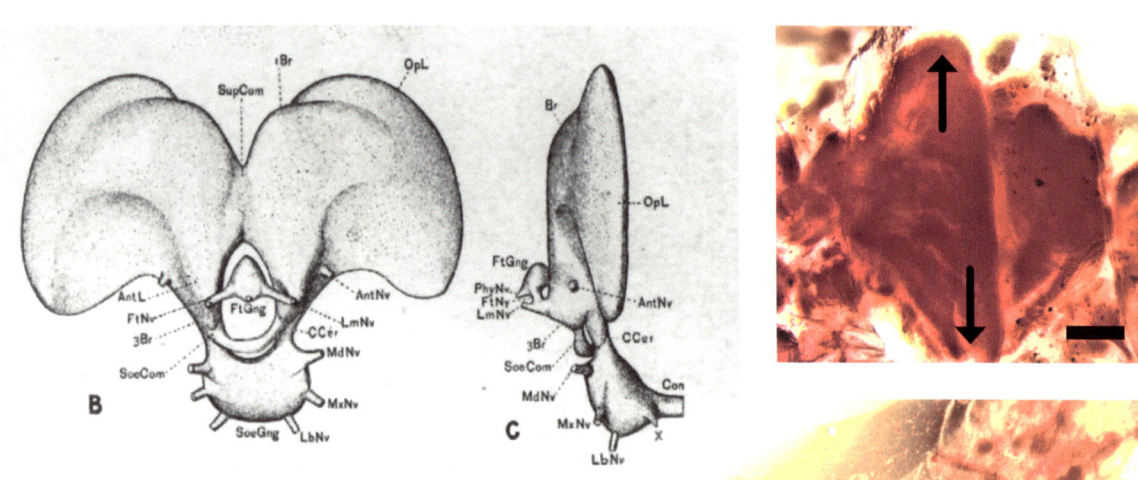

Fig 22 (above). Front and side view of the larval brain. Taken from Nelson.
Fig 23 (above right). Transparent image showing the brain.
Fig 24 (right). Close up view of the brain. Yellow arrow indicates areas develop into future optical lobes. Scale bar 100 microns.

Nerves

The nervous system of the mature larva is simple as compared with that of the adult and consists, in the mature larva, of a brain, a nerve chain comprising eleven ganglia joined by paired connective and a stomatogastric ganglion with its accompanying nerves.

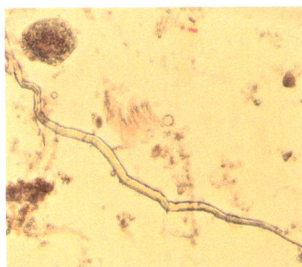

Fig 25. Nerve fibre. X 400 Magnification.

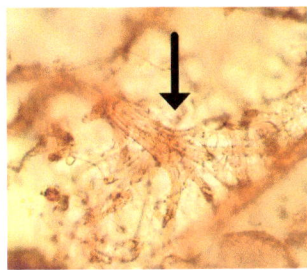

Fig 26. Nerves branching out from the ventral cord. X 100 Magnification.

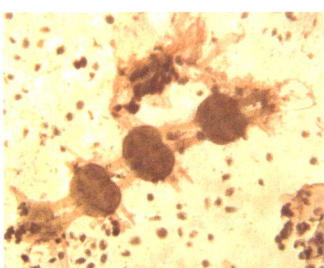

Fig 27. Five ganglia showing connecting paired, ventral nerve cords.

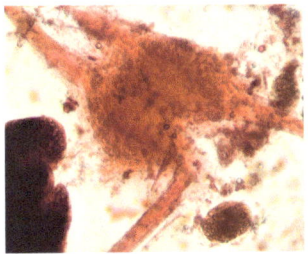

Fig 28. Close up of ganglion. X 200 Magnification.

The ventral cord has eleven ganglia that are connected to two separate wide nerves. All of these ganglia are to be found at the end of each segment except the eleventh, where it sits midway in the segment.

Each thoracic ganglion gives off two pairs of lateral nerves Also each abdominal ganglion has one pair of lateral nerves. The eighth abdominal ganglion has four pairs of nerves.

The eleven ganglia of the ventral cord are long in shape and are connected to one another by distinct parallel bands of nerve tissue. The three thoracic ganglia are the largest, those of the following seven abdominal segments being subequal; the eighth abdominal ganglion is, however, elongate and comprises three pairs of simple ganglia and the rudiment of a fourth. All of the abdominal ganglia are provided with well-developed lateral nerves, which divide into branches supplying the viscera and muscles.

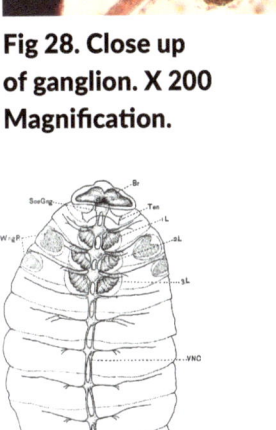

Fig 29. The larval nervous system, showing the six ganglia of the head all ready combined to form the brain and suboesophageal ganglion.

Oenocytes, excretory cells and fat bodies

There are three main types of cells, oenocytes and trophocytes, which are found in the larval fat body that loosely clump together, and the third type, urate cells that are responsible for storing nitrogenous waste.

The larval fat body occupies most of the space inside the body, except for internal channels, which permit circulation and are fixed by the trachea in places but float in the haemolymph. The transparency of the tegument gives the larva its creamy colour. Fat bodies are responsible for the intake, digestion and storage cells in the larva and make up about sixty per cent of the body weight. The fat cells do not store much fat until the larva reaches about three days old.

The fat body of insects has the same function as a liver in vertebrates.

The Oenocytes are insect cells responsible for lipid processing and detoxification. In honeybees, they are very conspicuous due to their large size and their great affinity when staining. They are found abundantly throughout the abdomen, always occurring singly, never in groups, and generally in more or less intimate contact with trophocytes.

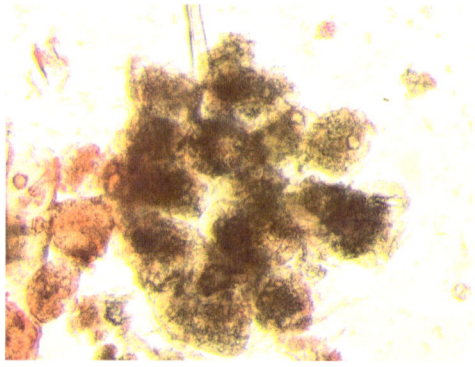

Fig 30. Oenocytes cell, stained red. Fat cells, brown. X 200 Magnification.

Excretory (Urate) cells are present in a limited number, scattered among the fat cells. Their role is to collect nitrogenous waste, primarily during larval development.

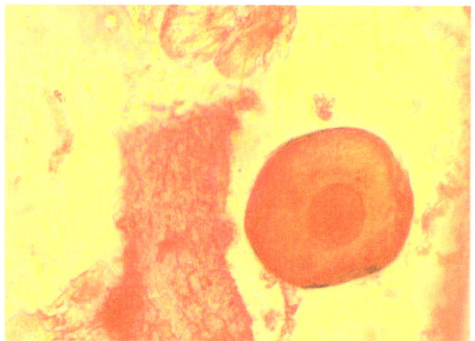

**Fig 31. Oenocytes cell.
X 400 Magnification.**

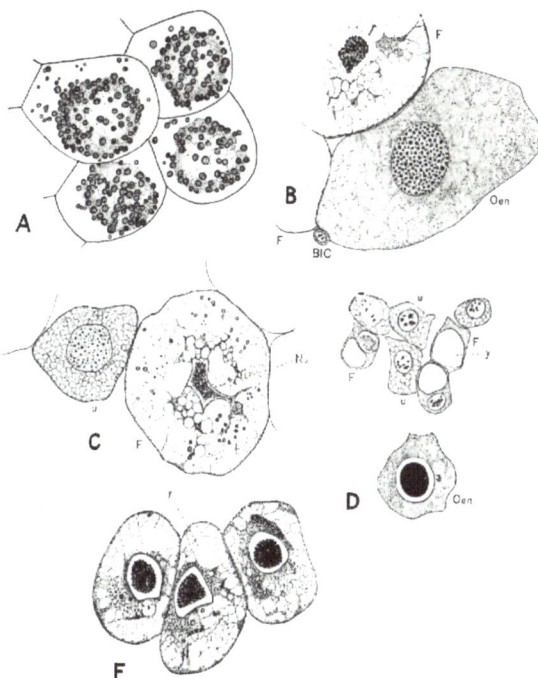

Fig 32. Taken from Nelson. 400 X Magnification.
A. Four fat cells from advanced worker larva, showing blacken fat globules.
B. Oenocytes *oen* and fat cells *f* mature larva.
C. Urate cells *u* and fat cell *f* from a mature larva.
D. Fat cells *f* oenocytes *oen* and urate cells *u*, about two-day-old larva.
E. Four fat cells from a three-day-old larva.
F. Three fat cells from a 3-4 days old larva.

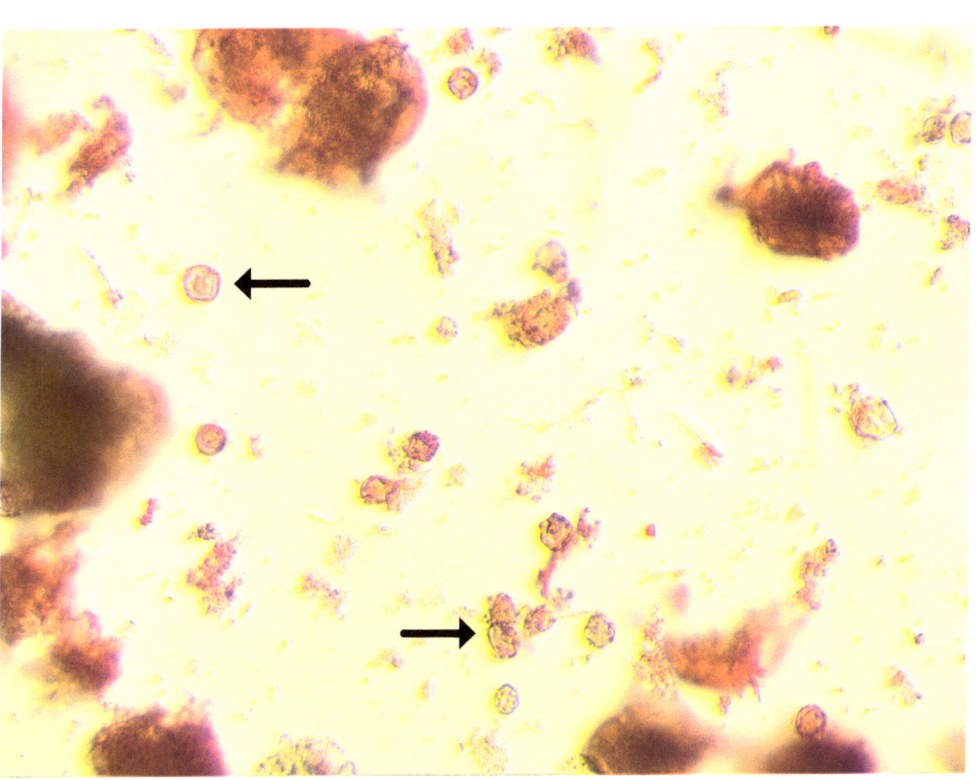

Fig. 33 Haemocytes cells are immune cells that circulate in the blood, the haemolymph; they help clear pathogens from the haemolymph. X 200 Magnification.

Digestion

The alimentary canal starts in the embryo as a small stomach formed in the middle of the body from the remains of the yolk of the egg. Small pits that occur at each end become the mouth and anus. The pits deepen and become inward growths, or stomodeum, the meaning of the nature of a mouth, and proctodeum, meaning anus. They meet the mesenteron, meaning the middle, forming a tube. These three regions then form the fore, mid and hindgut. Further growth occurs on the inner end of the hindgut becoming the Malpighian tubules.

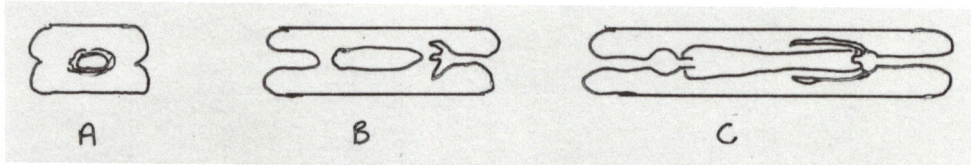

Fig 34. A. The midgut containing the remaining yolk, showing pits at each end. B. The stomodeum and proctodeum continue to grow inwards and the Malpighian tubules develop. C. The hatching stage, showing the hind and foregut connected but not open.

The alimentary canal is comprised of three sections, a short and relatively slender fore intestine, a large cylindrical mid-intestine and a hind-intestine. The fore intestine includes the mouth, pharynx and oesophagus. The mouth is a wide transverse slit passing immediately into the pharynx, which leads directly into the tubular oesophagus, which opens into the anterior end of the mid-intestine. There the wall of the oesophagus forms a fold that projects into the anterior end of the mid-intestine, developing into an oesophageal valve.

The mid-intestine is very capacious, cylindrical in form, about one-third of the diameter of the body and extends from the prothoracic to the ninth abdominal segment. Its walls are thick and composed of large cubical cells displaying a striated border.

The posterior end of the mid-intestine is completely closed and its extremity is not covered by the muscular coat. Four thin Malpighian tubules found pointing towards the head, increase as they are filled with waste matter.

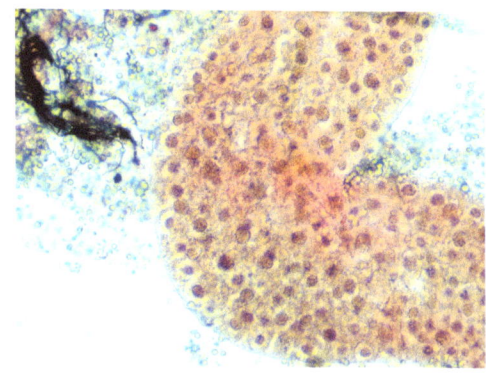

Fig 35 Stained intestine cells. X 200 Magnification.

The hind-intestine is a relatively slender tube, doubled on itself and of uniform diameter except at its anterior end, which is enlarged and closed anteriorly by a thin diaphragm-like membrane. Posteriorly the hind-intestine terminates in a slit-like anus.

The alimentary canal is the largest organ of the larva, reflecting the impact that, together with the mid intestine, or ventriculus, feeding has on development as well as the role of digestion.

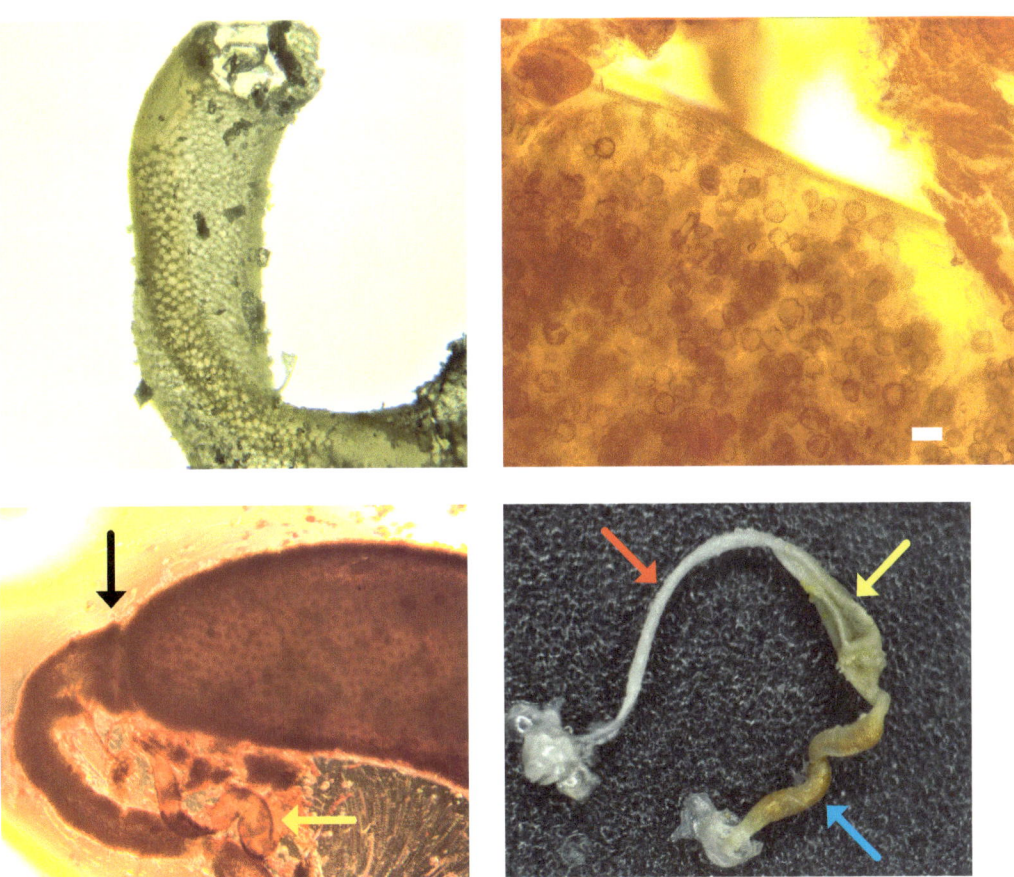

Clockwise from top left:
Fig 36. Empty mid intestine showing cell structure. X 100 Magnification.
Fig 37. Larva about eleven days old, showing pollen grains inside the gut. Scale bar 25 microns, 200 X Magnification.
Fig 38. Hind end of intestine showing valve area, black arrow, and the yellow arrow showing Malpighian tubule. X 40 Magnification.
Fig 39. Dissected alimentary canal of a larva, showing the three distinct areas, the foregut red arrow, midgut yellow arrow and hind intestine, blue arrow.

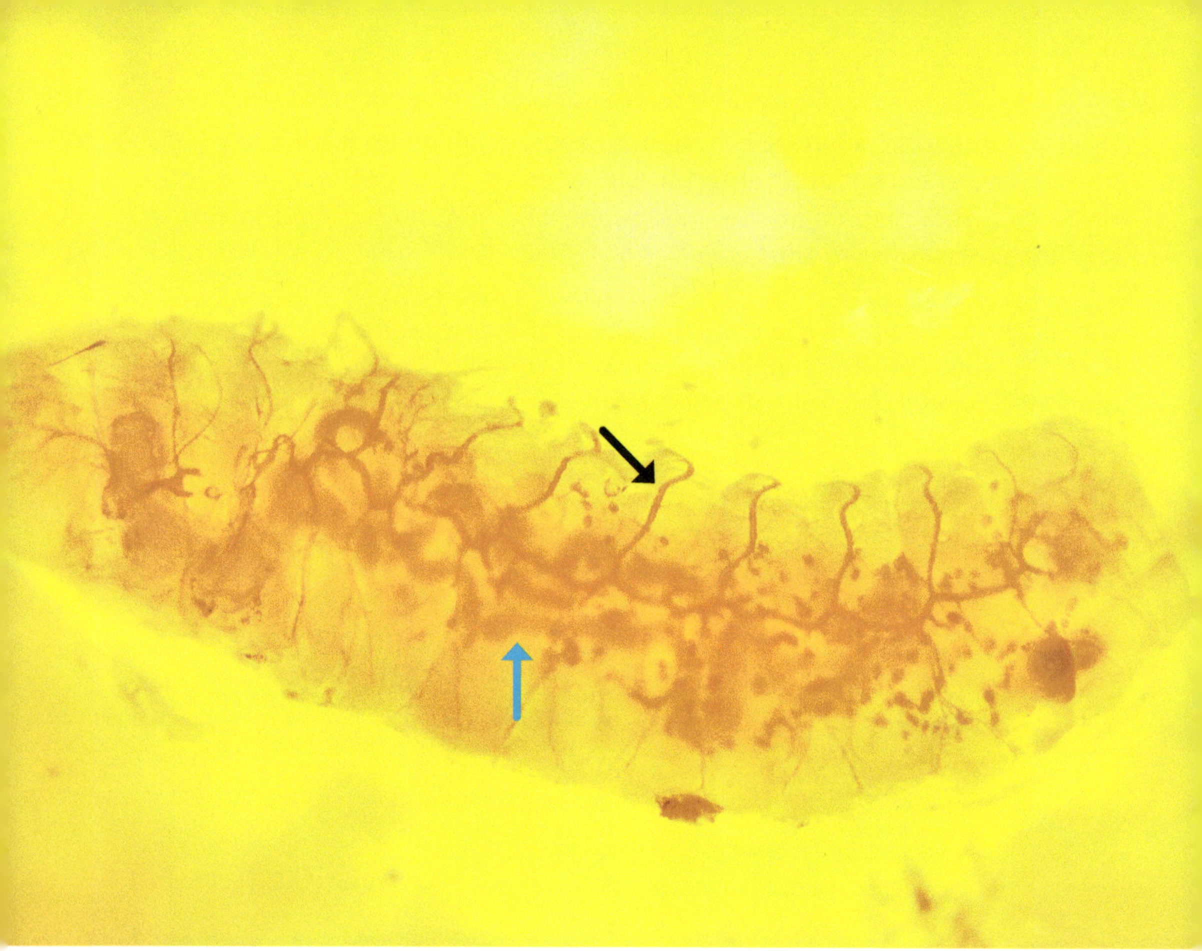

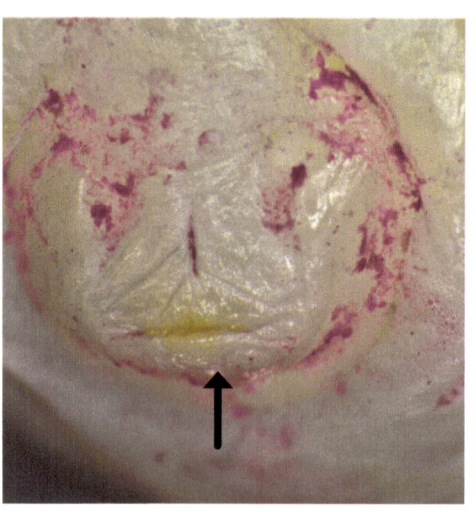
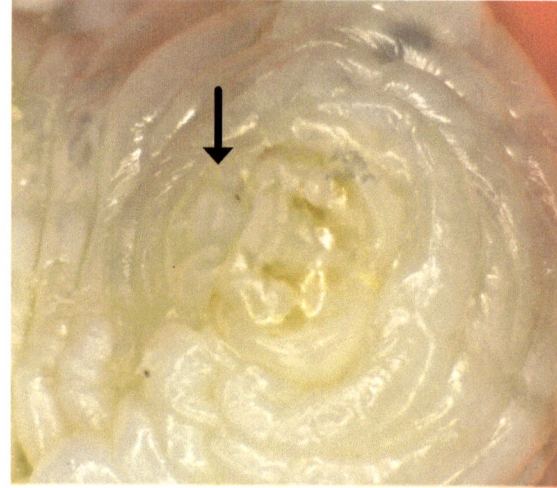

Fig 40 (top). Trachea, black arrow. Malpighian tube, blue arrow.
Fig 41 (bottom left). Stained and fixed larva, showing anus.
Fig 42 (bottom right). Rudiments of worker bee larva sting, known as gonapophyses, are sited behind the genital opening.

Malpighian tubules

There are four Malpighian tubules in the larva, two on each side between the lateral faces of the mid-intestine and the body wall. They pursue a winding course, forming numerous loops and folds, and extend from about the last thoracic to the ninth abdominal segment. In the newly hatched larva, the tubules are slender and of nearly uniform diameter throughout; in the mature larva their anterior ends only are slender, the tubules widening gradually reaching their maximum diameter, which is about one-half that of the mid-intestine, in the region of the seventh abdominal segment. The Malpighian tubules act like kidneys in mammals - they filter and store the accumulation of fluid urates, composed of nitrogenous waste, these products are absorbed from the haemolymph by Malpighian tubules and accumulated in their lumen (chamber) until the end of the larval feeding phase, day seven. Before the metamorphosis stage, they void the excrement stored in the intestine and the urates into the bottom of the brood cell were it is entangled with the silk cocoon. The larval Malpighian tubules then become inoperative and begin their degenerative process. At about two days after the fourth stage, the pupal moult, new adult tubules are being made from stem cells located at the insertion of the tubules into the digestive tract.

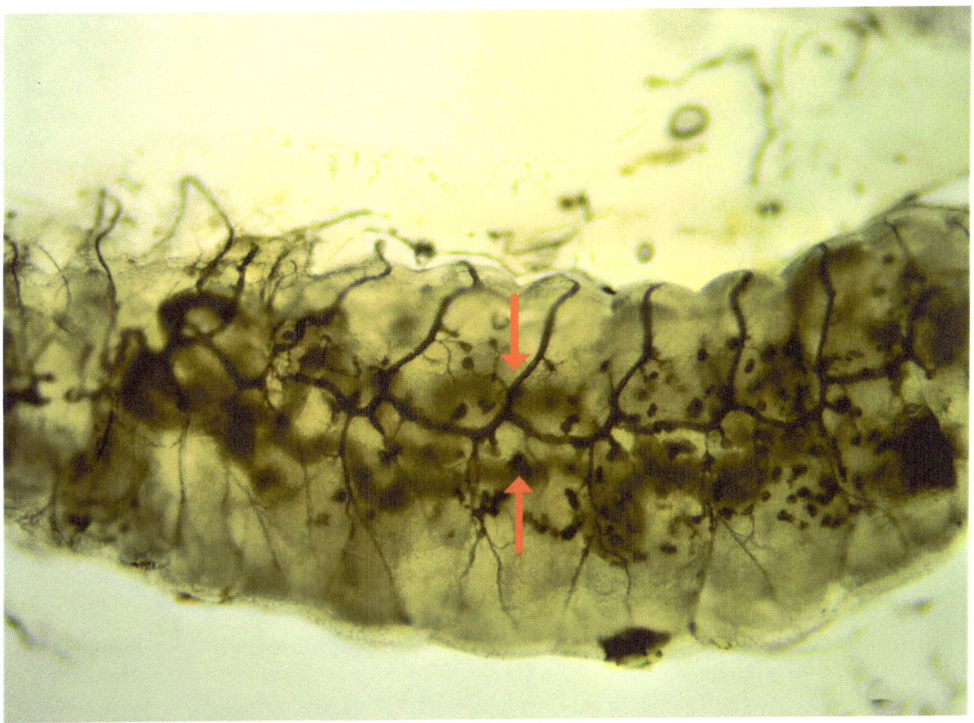

Fig 43. Malpighian tubes red arrows.

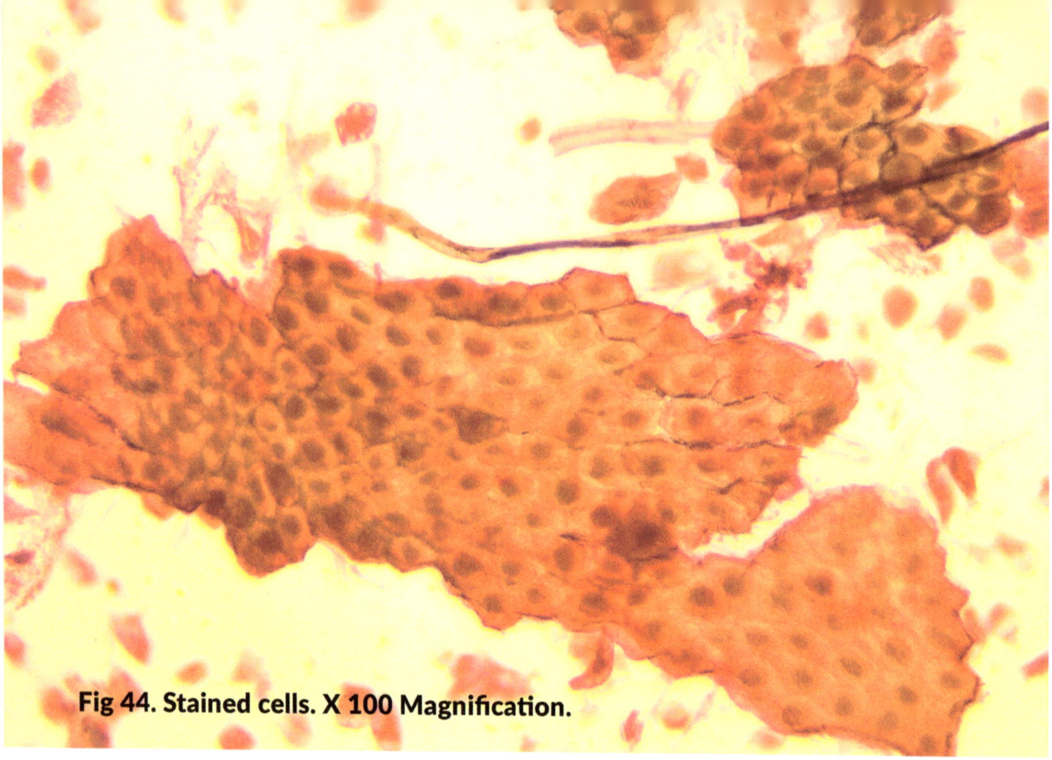

Fig 44. Stained cells. X 100 Magnification.

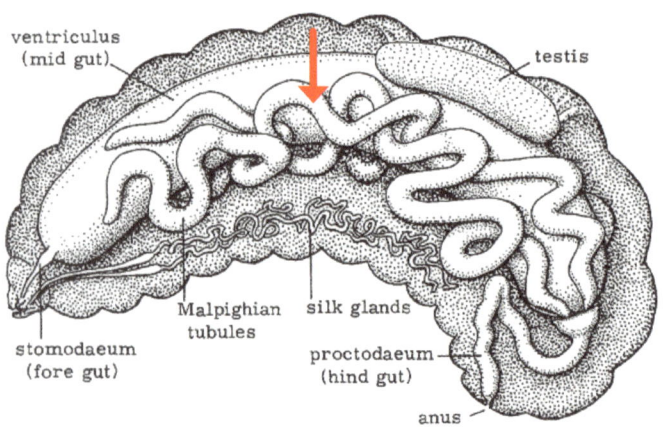

Fig 46 (above). Dissection of drone larva from the side, red arrow showing Malpighian tubules. Taken from Dade with kind permission of IBRA.

Fig 45. The walls of the Malpighian tubules consist of a single layer of epithelial cells. Taken from Nelson.

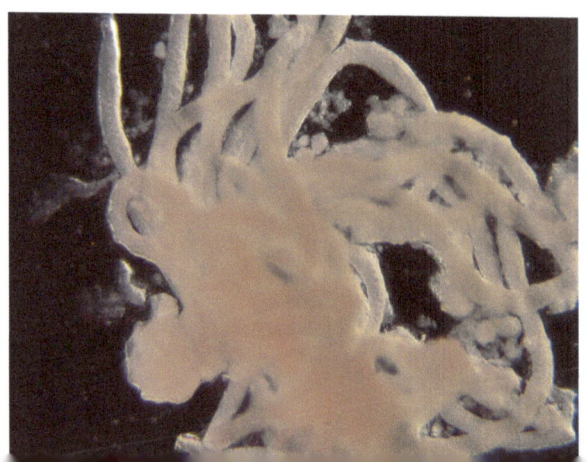

Fig 47 (right). Developing Malpighian tubes from the fourth, pupal moult of the larva. X 40 Magnification.

Tracheal system

There are ten pairs of external openings of the tracheal system, named spiracles that are arranged on each side of the body, midway between the ventral and dorsal mid-lines. Each spiracle is a minute, round aperture situated in the middle of a small circular elevation of the surface, the stigma, meaning a mark or sign. On all of the segments bearing spiracles, a shallow linear depression or suture arises in the neighbourhood of each spiracle. The divisions marking each segment circle the body is related to the internal muscle structure.

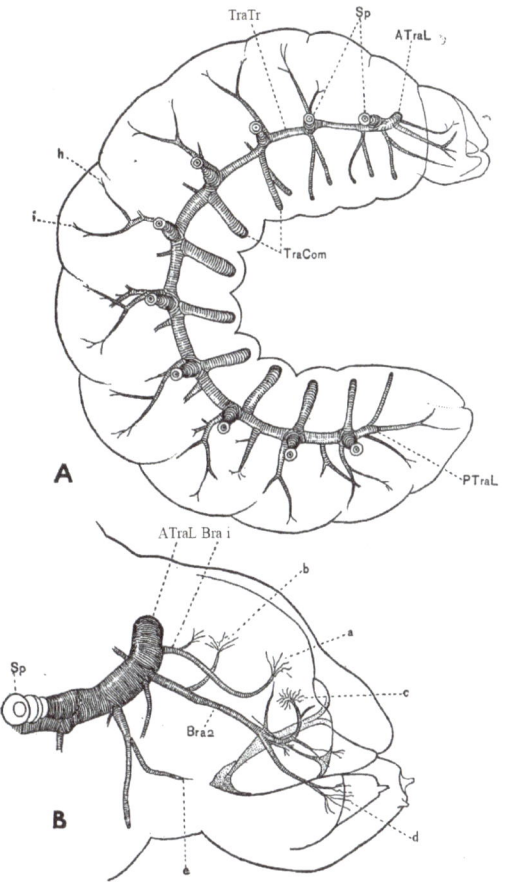

Fig 48 (above). The tracheae are simple in structure, being merely thin-walled tubes composed of small flat epithelial cells and lined with chitinous bands, that thickened to form fine transverse spirally wound threads, the Taenidia. The size of the Taenidia is at least approximately proportional to the size of the trachea of which they form a part. Taken from Nelson.

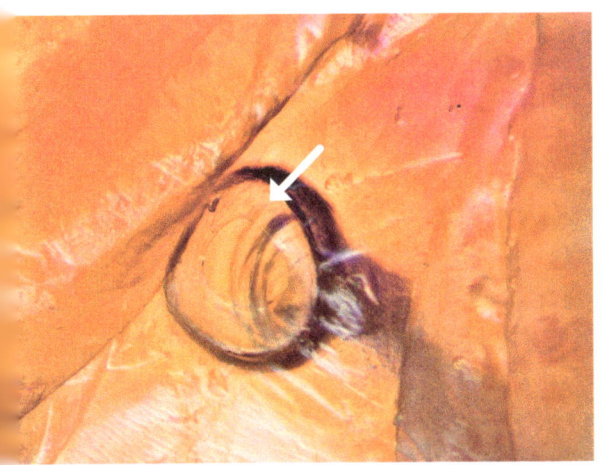 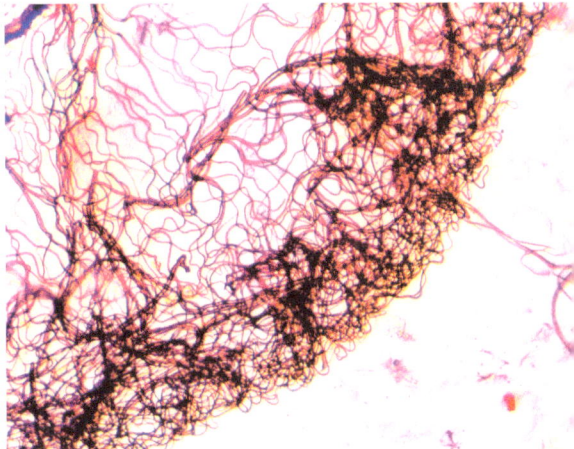

Fig 49 (left). Spiracles, showing membrane. X 100 Magnification.
Fig 50 (right). Tracheoles were taken from the larval epidermis. X 40 Magnification.

The spiracles are connected by short branches, the spiracular branches, to the tracheal trunk of the corresponding side. Each of these trunks traverses the body cavity, about midway between the hypodermis and mid-intestine. The tracheal trunk of a mature larva has an inside diameter of about 0.1 mm. in its posterior; in the three anterior segments, its diameter is reduced to about 0.05 mm. The anterior ends of the tracheal trunks of opposite sides meet to form a loop. In the neck region above the oesophagus, there is a similar loop.

Throughout their length, the tracheal trunks give off numerous branches that supply the various regions of the body. Three branches on each side supply the head. Segmentally arranged branches surround the longitudinal tracheal trunks supplying the muscles, heart and viscera. The tracheae terminate in typical tracheal end cells from which the smaller tracheoles arise.

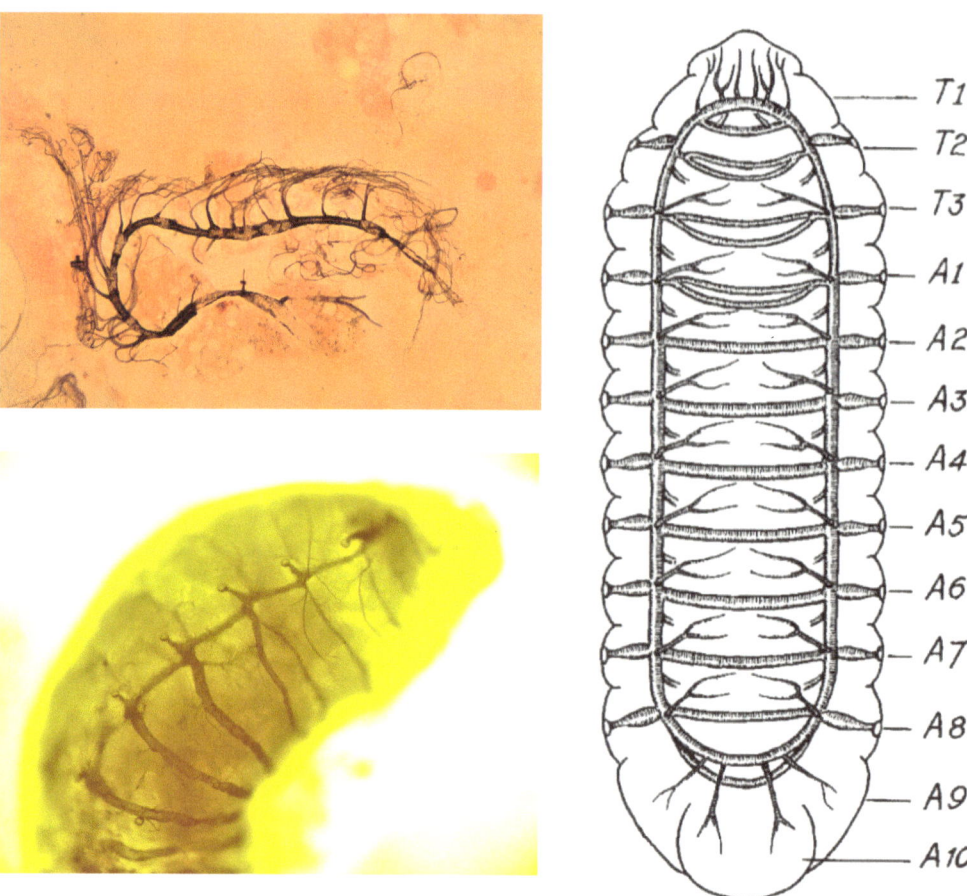

Fig 51 (top). Fine trachea structures found throughout the larval body. X 200 Magnification.
Fig 52 (above). Showing Spiracles and network of supporting trachea.
Fig 53 (above right). The larval tracheal system from above showing a looped system. Dade Fig 21.

Muscles

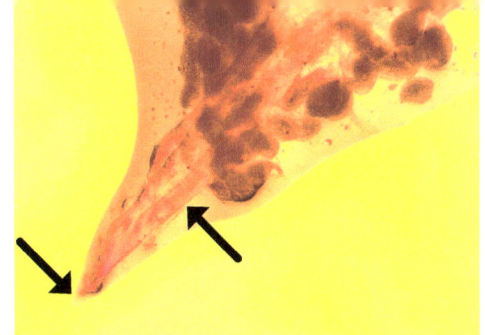

Since the body wall is not rigid, the arrangement of the muscles is responsible for the external contour of the body. The larva has a network of muscles that allow some movement of the body and the mouth.

There are fewer muscles in the thorax and the tenth segment has a few retractor muscles that act on the anus.

The muscle structures include longitude and oblique muscles that help hold the body shape and mesh into the body wall. The tergosternal muscles aid in breathing.

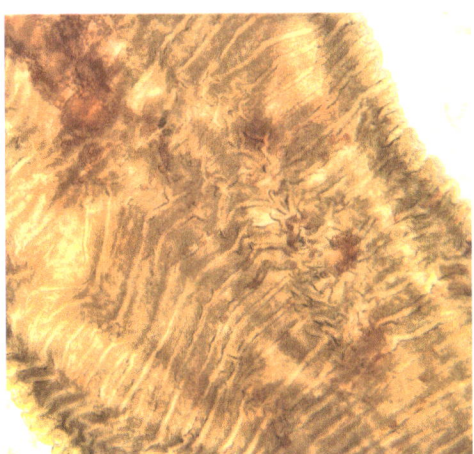

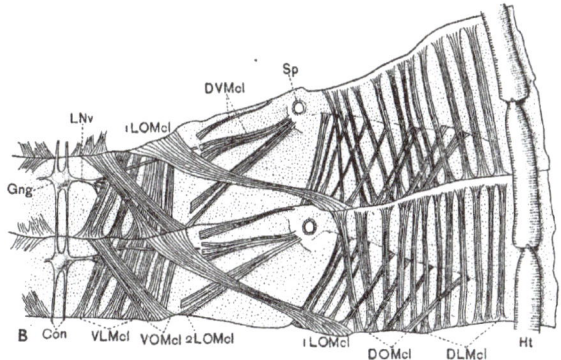

Fig 54 (top right). Section showing the muscles in the larval mandible. Note spine at the end, called a palpus, Latin for a feeler, which is used as a sensing organ. It is thought that they can taste sugar provided from the brood food and salts supplied from pollen.
Fig 55 (middle right). Dorsal lateral muscles.
Fig 56 (bottom right). The inner surface of the fourth and fifth thoracic segment, showing musculature X 10 magnification. Taken from Nelson.
Fig 57 (bottom left). External view of the muscle layer beneath the skin of a larva.

Heart and blood

The heart consists essentially of a slender thin-walled tube situated in the mid-line close beneath the dorsal hypodermis. It measures 0.25 mm at its widest at the posterior end and gradually diminished in size up to the anterior border of the second trunk (the mesothoracic) segment.

The aorta is not strictly tubular but is open on the ventral, underside, having in transverse section the form of an inverted letter U, the free edges hanging down on each side of the oesophagus and becoming clothed on the exterior with a layer of tracheoles.

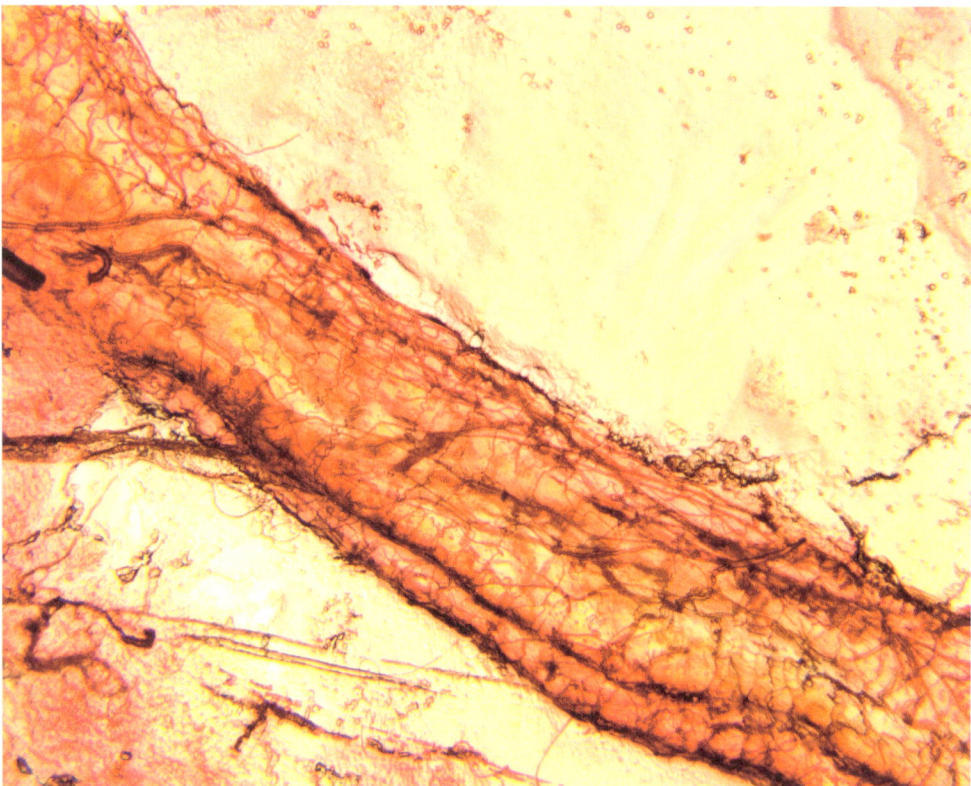

Fig 58. Supporting a network of trachea around the heart.

At the posterior side of the brain the aorta becomes reduced in its dorso-ventral diameter to enter the narrow cleft between the oesophagus and the upper ends of the crura cerebri and finally, blindly, terminates at the anterior face of the brain. In life, the heart is transparent, (meaning obscure) and since it is bounded on either side by the relatively opaque white fat cells, it produced externally the appearance of a dark band along the dorsal midline of the larva. At the middle of trunk segments two to eleven inclusive, the heart is sharply constricted and is thus divided into eleven chambers. These constrictions, however, do not affect the dorsal and ventral walls of the heart, but only the lateral walls, which are indented by a series of pairs of opposite V-shaped indentations with the open end slightly elongated. At the bottom of each indentation is a linear slit; these slits constitute the Ostia, meaning a door. The heart walls bordering on the Ostia form valve-like flaps projecting inward which allow a free inrush of blood during diastole, filling, but which automatically close the Ostia during systole. The posterior pair of flaps project inward far enough also to close the posterior ends of the heart chambers during systole, and the pumping motion prevents a backward flow of blood. The wall of the heart is exceedingly thin and is composed of a double row of cells.

The heart is similar to the adult bee. It extends from near the end of the larva A9 where it is a blind end to just behind the head and is divided into eleven chambers by ten pairs of Ostia. These are visible through the skin on the larval back.

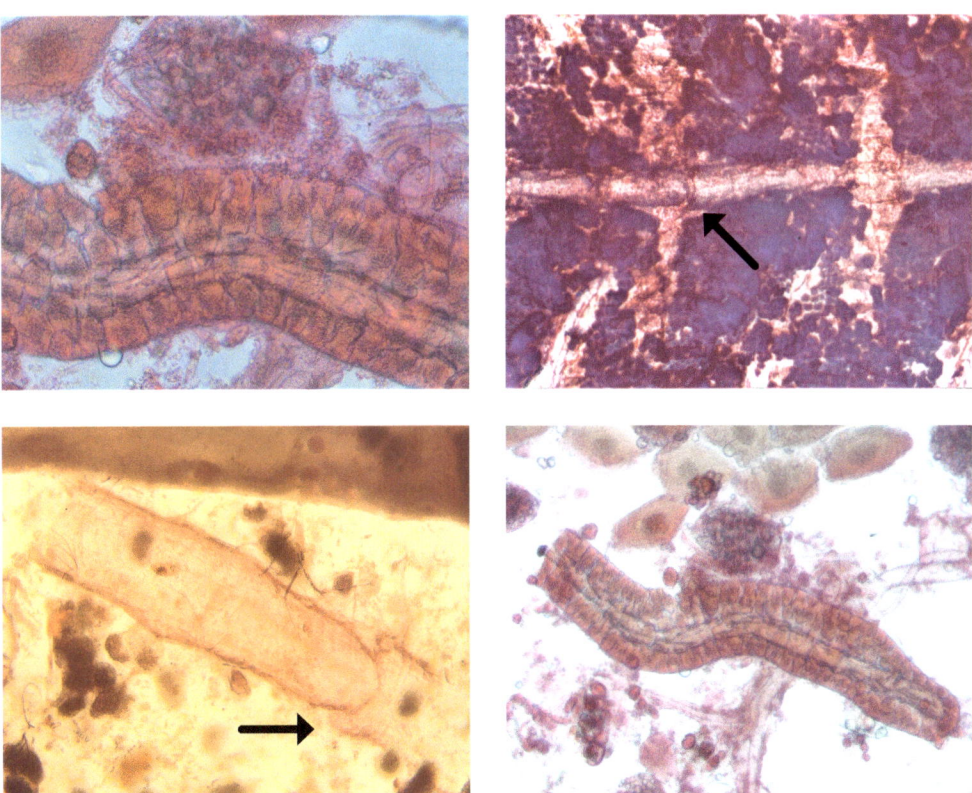

Fig 59 (top left). Heart showing supporting cells.
Fig 60, 61, 62 (clockwise from top right). Ostia of heart. Area of heart, surrounded by cells.

Externally the heart is clothed with a network of branching minute connective tissue cells. A dorsal diaphragm is present which is especially well developed in the fourth to the ninth abdominal segment.

The blood cells of the larva are all of one kind, of minute size and ellipsoid form. Many are found in the stages of division, indicating that this is the chief, if not the only, method of increase, since no blood-forming tissue is found.

Dorsal diaphragm

Structurally the dorsal diaphragm consists of two very delicate membranes, having the appearance of basement membranes and possibly being chitinous. These membranes are attached to the ventral wall of the heart in the mid-line.

Ventral diaphragm

The ventral diaphragm is well developed in newly hatched larvae and the imago and forms a continuous sheet made up of transversely arranged muscle fibres overarching the ventral nerve cord and partitioning off a ventral (perineural) sinus. In older larvae, it becomes merely a vestigial structure confined to the abdominal segments.

Ovaries

The reproduction organs are present in the larva in rudimentary form. The ovaries are a pair of bonded bodies near the back of the fifth abdominal section.

The rudiments of ovaries in the queen larva are much larger than in the worker larva, showing that their development is greatly accelerated during the later larval stages. However, over 90% of the ovariole primordia that were formed during the embryonic and early larval stages degenerate in the last larval Instar by massive programmed cell death. Their structure is similar to that of the worker larva except that the presumptive rudiments of ovarian tubes are both more numerous and longer.

Testes

The rudiments of testes in the drone larva are relatively enormous, lying in the fourth, fifth and sixth abdominal segments, and are composed of numerous transversely arranged strands of cells united by connective tissue. The vasa deferentia are present. They run from the dorsal testes ventrally around the large intestine to the gonadal imaginal discs.

Drone production and development

The strain of the bee will influence the number of drones laid and when. The Italian queens, the most popular race in Britain, produce drones early in the season.

Drones are reared in larger cylindrical cells (6.3 mm diameter) than workers (5.3mm) on a 9-13° angle from the horizontal and have raised domed cappings, all within a supporting hexagonal comb structure.

They are mainly found on the outside of the brood nest, often at the bottom of the frame. Overall, typically about 15% of the brood nest comprises drone cells. Unfortunately for the drones, varroa mites prefer drone comb in which to reproduce; they have more room to develop and more importantly, extra time to reach maturity-24 days compared to 21 days for workers.

Fig 63. Lateral view: Late-stage of drone imago, a day or so away from exiting the cell.

Drone rearing starts at the beginning of the foraging season when nectar and pollen become plentiful.

In temperate climates, drone eggs are therefore laid from late spring to late summer and peak drone population occurs during swarm season. All colonies in the same geographic area are usually synchronised to start drone rearing at the same time but drone production tends to start earlier in colonies with large worker numbers.

Colonies with larger food reserves produce more drones. In contrast, drone comb is not constructed in colonies with less than 4,000 workers and drones are not produced at all in small colonies or those with a high disease burden or when environmental conditions remain persistently unfavourable.

To ensure a sufficiently large drone population has been established in the locality before the first virgin queen mating flights, drone-rearing starts about three to four weeks before queen rearing. However, other than drone rearing preceding swarming, the two activities are not closely correlated within individual colonies.

Whilst swarming colonies tend to produce more drones, the presence of drones per se is not predictive of swarming, although their absence can be a contra-indicator. Successful drone rearing is draining of colony resources and rests upon the maintenance of exacting nutritional and brood nest conditions.

The presence of a drone brood slows down the rate of further drone egg-laying by the queen. Furthermore, she will not lay in drone cells in very poor weather.

Drone eggs and larvae are cannibalised by workers whenever weather conditions or colony nutritional reserves are very poor. No eggs at all are laid in drone cells from late summer onwards.

In the late summer from the end of July to the beginning of August, drones are no longer fed by workers and they are evicted from the nest to die. This eviction is delayed if colonies receive autumn feed and it does not happen at all in queenless colonies or failing queen colonies.

Although about 5,000 eggs will be laid in these drone cells annually, only about 2,000 drones will develop into sexually mature bees, a 40 % success rate.

As previously stated, drones take the longest time to develop of the three castes, 24 days. A queen takes 16 days, workers 21 days.

One reason for the time difference may be that the drone's body is given over to the development of the testes. Primordial germ cells develop in drone larvae the moment they hatch from the egg. Spermatogenesis takes place rapidly, resulting in the production of spherical sperm cells that have a head but lack tails.

Fig 64. Drone raised capping.

All sperm cells have been synthesized by the sixth day of larval development. They mature, but they do not multiply during development and continue to mature into functioning sperm for about twelve to fourteen days after emerging from the comb. At this time, the drone becomes sexually mature. Sperm degrades during the lifetime of the drone so peak fertility is not maintained.

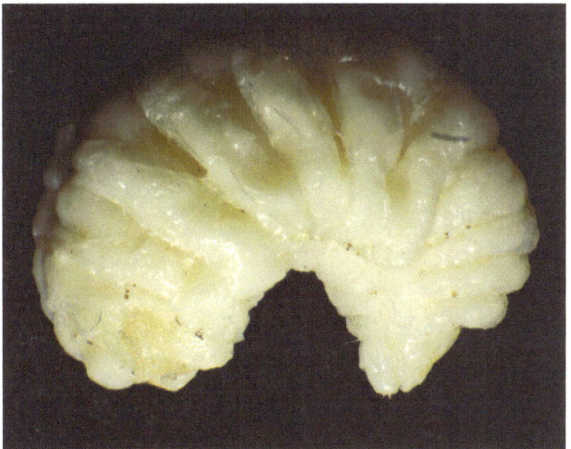

Fig 65. Drone larva about seven to eight days old.

Sperm will then be stored in the drone's seminal vesicle ready for mating and the testes will atrophy.

Drone larvae need considerably more food than worker larvae and receive considerable quantities of brood jelly, which contains a wider range of proteins. Older larvae receive pollen and nectar in their diet.

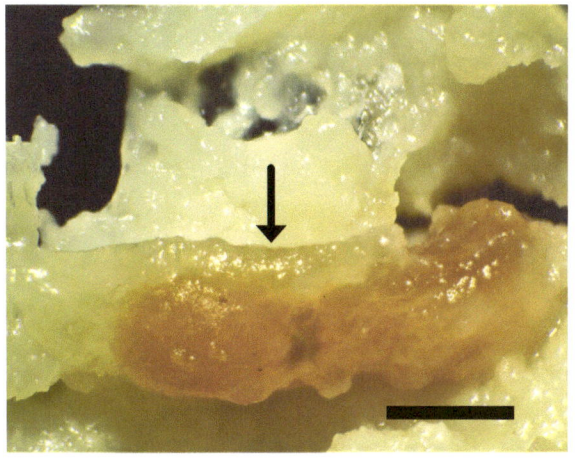

Fig 66. Same drone larva as drawing 65, showing testis. Scale bar 1 mm.

Workers are more attentive to the thermoregulation of drone brood than worker brood. Primordial germ cells develop in drone larvae the moment they hatch from the egg.

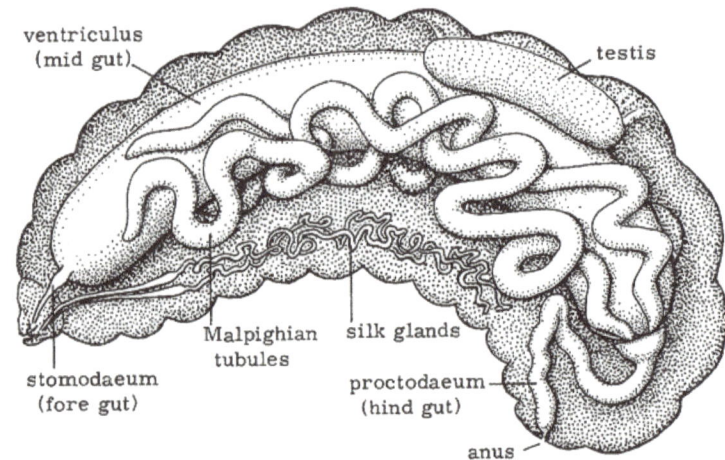

Fig 67. Sagittal section of larva from the lateral aspect. Taken from Dade by kind permission of IBRA.

It is also at this hatching time that the sperm cells transform into spermatozoa with heads and tails.

Whilst initial development of the endophallus also starts during pupation and the sclerotized plates are completely synthesized in this developmental phase, a considerable amount of the reproductive apparatus is only made post-eclosion, i.e. after the drones emerge from their sealed cell for the first time, day 24 of their lives.

Fig 68. Drone eyes and flight muscles develop during pupation.

Fig 69. Some drones emerging from cells note raised drone capping on the right.

Sexually immature drones

For the first eight days post emerging, young drones are fed a combination of brood food, honey and pollen by worker bees and they remain in the warmth of the brood area whilst undergoing their second phase of maturation, which is both physical and sexual.

It is during this time that much of the reproductive apparatus first develops. Sperm only starts to migrate to the seminal vesicles on day three post emerging.

When fully mature, the endophallus and related glands are tightly packed within the abdomen to conserve space.

Six to eight days post emerging, drones start making their first cleansing flights, which last about two and three minutes. At about this time, they move to the periphery of the hive and start to feed themselves.

Fig 70. Drone pupae displaying eye colour. Photo Copyright. Dr Michel Asperges University.

Silk wrappings make strong bonds

Silkworms are the only insects that have been domesticated by humans for their silk. We all know spiders spin silk to make webs to catch food, and moths to make their cocoons that can be found suspended from twigs. Doctor Sutherland and her team from the University of Waikato, New Zealand have already identified the honeybee silk genes and now they have identified and sequenced the silk genes of bumblebees, bulldog ants and weaver ants, and compared these to honeybee silk genes. This has enabled them to identify the essential design elements for the assembly and function of double-coiled silks. Bees and ants produce high-performance silk and, although the silks in all these species are produced by the larvae and by the same glands, they use them differently. Honeybee larvae produce silk to reinforce the wax cells in which they pupate; bulldog ant larvae spin solitary cocoons for protection during pupation; bumblebee larvae spin cocoons within wax nests and the cocoons are reused to store pollen and honey; weaver ants use their larvae as 'tools' to fasten fresh plant leaves together to form large communal nests.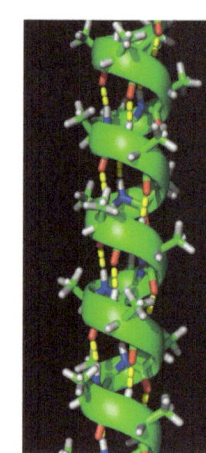

Most people are unaware that bees and ants produce silk and that its molecular structure is very different from that of the large protein, sheet structure of moth and spider silk. The cocoon and nest silk of honeybees consist of two separate coiled threads that combine, (known as an Alpha helix - a protein structural arrangement where multiple helices wind around each other, see diagram.) This structure produces a lightweight, very tough silk.

Natural combs are used to store honey and pollen and to rear the brood. The comb cells are constructed from wax secreted by young worker bees. The queen deposits the eggs in the cells that then develop into larvae, which are fed by nurse worker bees. After about six days, they have shed their cuticle four times and have grown into large larvae filling the cell completely. Feeding stops and the worker bees then cap the cells off.

A researcher named Jay reported that silk is generated from the labial gland as the larva perform random head movements in all directions whilst moving in slow somersaults within the cell. This behaviour may last up to 48 hours; it ensures that the final cocoon will be a randomised and mechanically

Fig 71 (top right). Alpha helix. Image by Bikadi. CC Attribution-Share Alike license.

planar isotropic structure. Meaning it has the same structure in all directions. During this time, the joint between the midgut and hindgut opens and the Malpighian tubules open disgorging all of their waste into the cell to be mixed up with the silk. The larva starts the fifth moult about two days after capping, the pupal moult. At the end of the pupal phase, the adult bee is formed inside the thin pupal cuticle, which is cast shortly before emergence and remnants of it can still be seen clinging to the young bee's body.

Silk proteins are produced by modified salivary glands. These long, kinked glands extend through most of the larva's body. They unite in a common duct and secrete into the gland lumen, a chamber where they accumulate to high concentrations during the final larval Instar. The gland opens in the spinneret on the labium. The glands become the thoracic salivary glands in the adult bee. There are five stages of moulting due to their rapid growth, each called an Instar.

The diameter of the silk thread is about three microns; they are embedded in the wax in a mostly random pattern with an occasional regular arrangement. The mass of silk in cocoons in the walls of a one-year-old honeycomb is about 33%; the rest is wax, excrement and secretions from the Malpighian tubules.

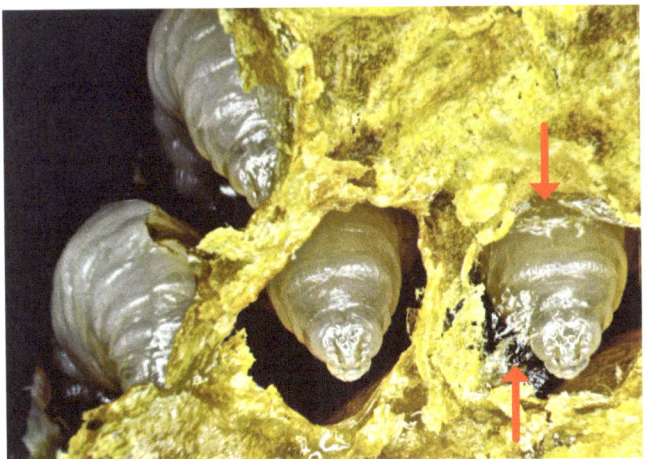

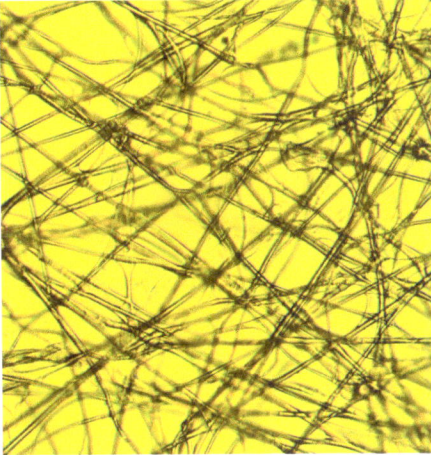

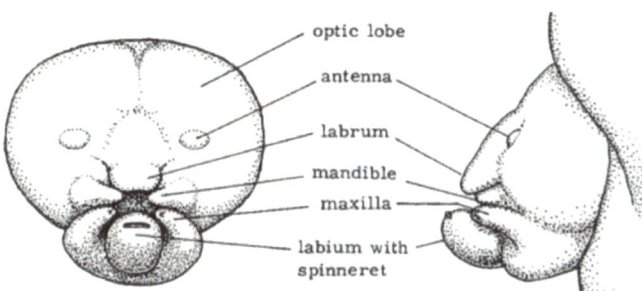

Fig 74. Head of a larva showing spinneret, taken from Dade by kind permission of IBRA.

Fig 72 (above left). Comb showing larval silk, red arrows, caps have been removed. X 40 Magnification.
Fig 73 (above right). Silk removed and cleaned from the capping of a sealed worker brood. X 100 Magnification.

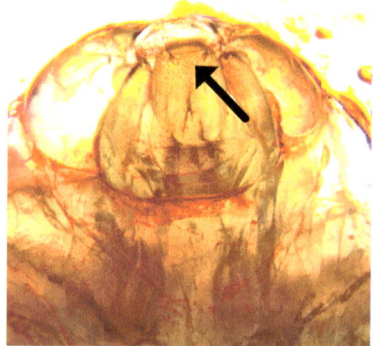 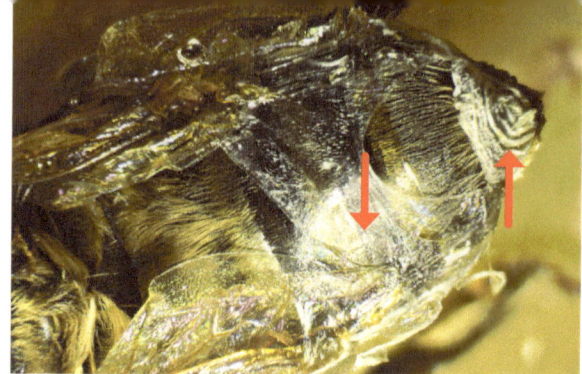

Fig 75. Head of a drone larval showing labium with spinneret X 100 Magnification.

Fig 76. Remaining pupal cuticle from the final moult, on a newly hatched drone.

After the pupae have metamorphosed into bees and left the cells, the worker bees clean any old shedded larval skin and cover the remaining silk with wax. Thus, the comb becomes a composite material with usage and gains some strength and rigidity. The cell is slightly reduced down in size as the walls thicken after each successive larva.

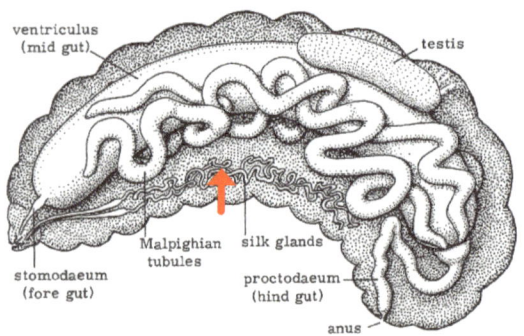 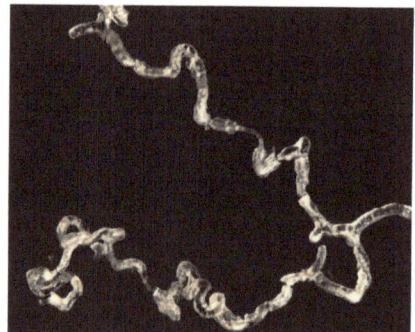

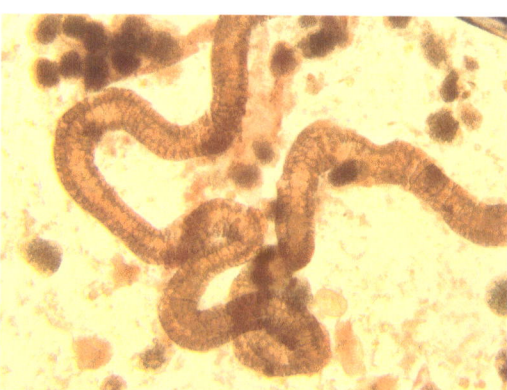 

Fig 78. Section through the larva showing silk glands, taken from Dade by kind permission of IBRA.
Fig 79A. Silk gland dissected out from larva showing lumen joint. X 40 Magnification.
Fig 79B. Convoluted silk gland. X 200 Magnification.
Fig 80. Capping removed to expose silk with bees inside of cells.

Pupal stage

Once the larval cell is capped off, the larva will change position constantly to spin a web of silk forming a cocoon around its body, ending face upwards. The larval cuticle is not shed for another four days when it becomes the prepupa and the fifth moult then takes place and the pupa is recognisable, although the internal organs have yet to form.

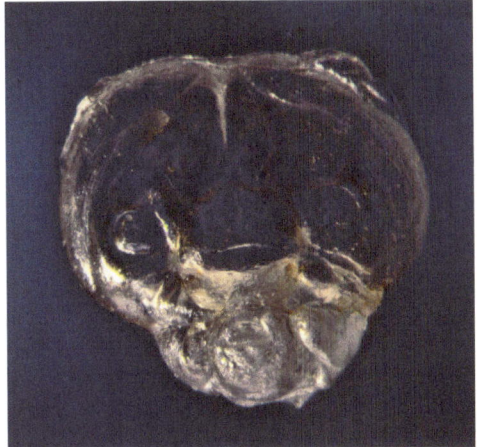

Fig 81. Larva head capsule shed at the fifth moult.

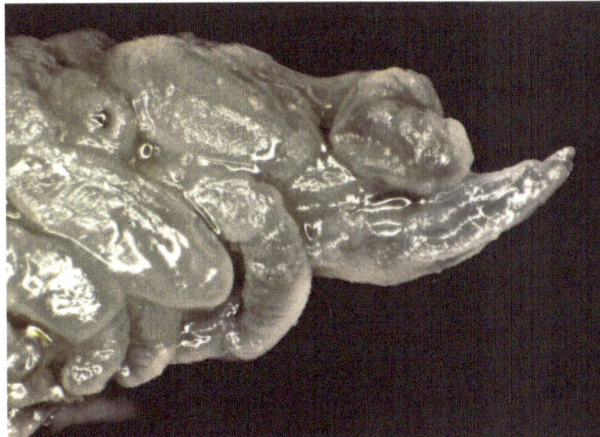

Fig 82. Prepupal stage, with larval cuticle, removed showing leg and mouth development.

From about day fourteen for a worker bee, the eyes start to colour, starting at a pale yellow to pink, purple and finally a brown-black.

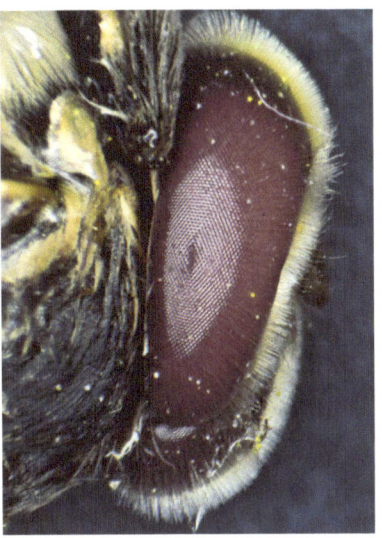

Fig 83 (above). Start of eye development. Drone bees.
Fig 84 (right). Purple stage of eye colour.

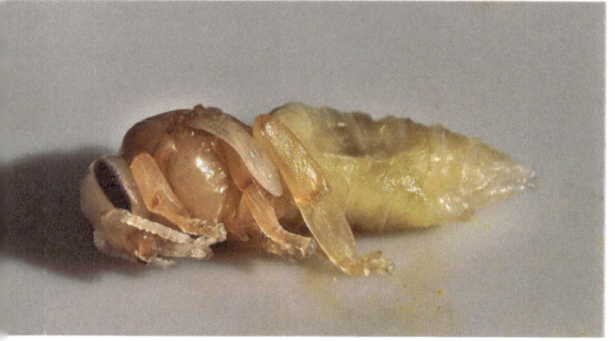
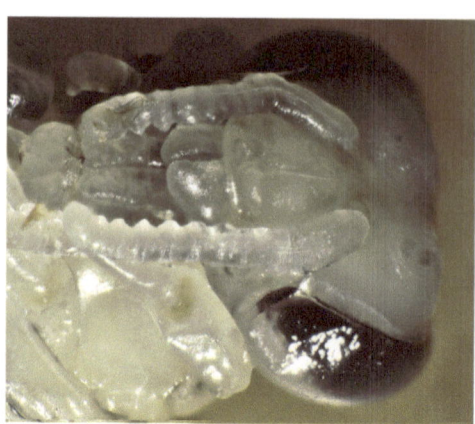

Fig 85 (above). Queen larva, note longer abdomen.
Fig 86 (right). Pupal head about day 16.

The ninth day sees all of the adult internal organs developed. The cuticle starts to darken from day 18 and continues when the final sixth moult takes place on day twenty for the worker bee.

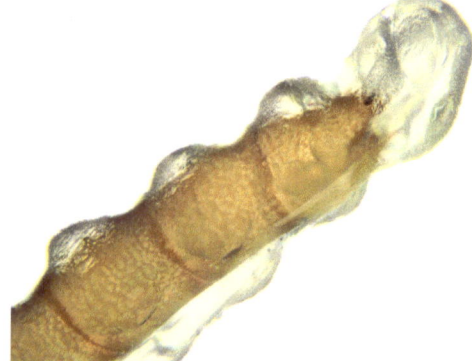

Fig 87. Close up of head about day 18.

Fig 88. Antenna, close up from the same bee above shown the last moult about to happen.

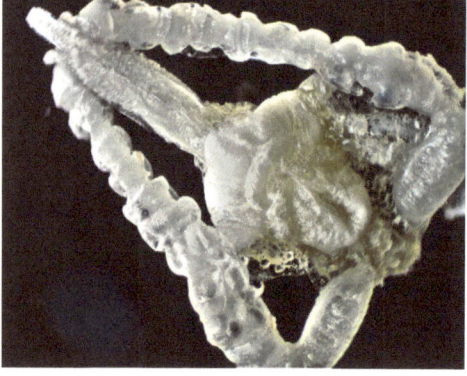
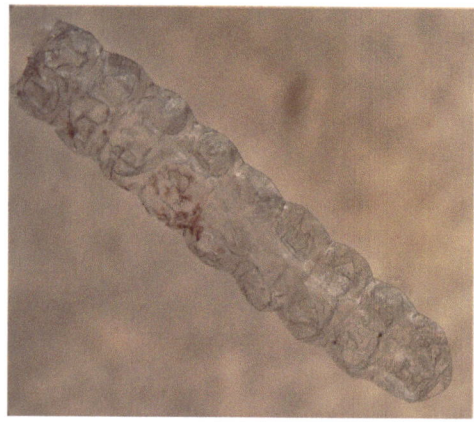

Fig 89 (above). Resulting final sixth moulted cuticle.
Fig 90 (above right). Antenna discarded cuticle from the final moult.

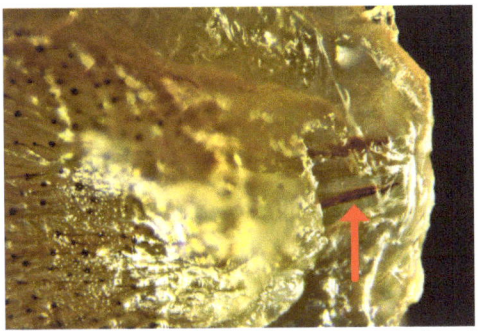
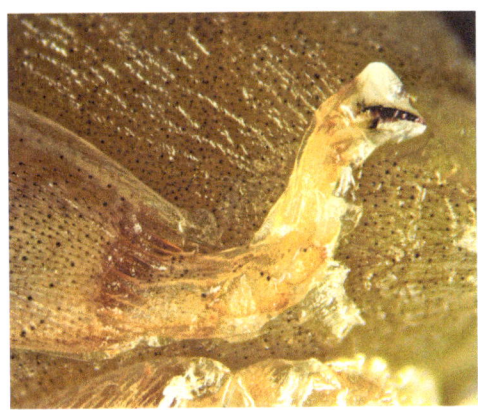

Fig 91 (above). Developing sting.
Fig 92 (right). The third leg development.

The process of development of a new bee takes place over about ten days once the cell is capped. The heart and the tracheal system are retained. The Malpighian tubules are dissolved and a new set of over one hundred are made. The digestion system is broken down and rebuilt. The brain and nervous system are kept but the ganglia merge in some places and new nerves are made with the developing body parts. Some of the bodily parts are lost in the process, e.g. the external muscle nerves surrounding the larva cuticle dissolve.

The sexual development starts in the larva and continues to develop throughout the pupal stage, in the case of the drone sperm continue to develop in the first few days as an adult.

Fig 93. Worker about day twenty, still to have the final moult.

Fig 94 (above). Wings are formed but need to be expanded and dried one emerged from the cell.

Fig 95 (left). A new worker bee hatching.

Fig 96 (left). Newly emerged drone bee, drying off.

Adult bees

Fig 97. An older drone, captured when returning to the hive on a sunny afternoon. Note the loss of hairs on the top of the thorax. The abdomen is more curved than the workers are.

Respiration

Respiration (noun.)

Late 14c, from Latin respirationem (nominative *respiratio***) "breathing, respiration," noun of action from the past participle stem of** *respirare* **"breathe again; breathe in and out," from** *re-* **"again" +** *spirare* **"to breathe".**

First, shave your bee

Not the easiest thing to do under the dissecting microscope! However, it is necessary to shave the bee first to see their external breathing orifices. Honeybees have external respiratory openings called spiracles, of which there are three pairs on the thorax and seven pairs on the abdomen; the last one is hidden within the sting chamber. They regulate air to every organ in the bee's body. The spiracles access the trachea arms, which inflate the nearby air sacs. Each spiracle has an internal valve to control the flow of air and has a muscle to open and close it. The valve in the first spiracle cannot close fully, which can be problematic in that it is here that the acarine mites enter the bee's trachea. Once inside they pierce the trachea causing disease as they feed on the bee's haemolymph. Large numbers can block the main airways, although this does not seem to impede the bee when flying.

Another reason for keeping the spiracles closed at rest is to prevent internal dehydration as the tracheal system is kept moist; bees can tolerate high concentrations of CO_2 and only need to breathe at rest when this gets too high. A build-up of carbon dioxide stimulates the nerves in the ganglia, which control the muscles that stretch and contract the membrane linking the sternum and tergum. This leads to contractions of the muscles forcing air (CO_2) out via the spiracles; releasing the muscles causes a vacuum allowing fresh air in. A similar action occurs longitudinally in the abdomen as well as in the thorax.

As a comparison to the human respiratory system, both have an area that expands for air intake. However, in humans oxygen must be bound to the haemoglobin molecule in the blood to be transported; this is known as a wet system. CO_2 also leaves the tissues in an aqueous solution via the bloodstream to be expelled via the lungs when we breathe out. In the bee, the air is transported as gas to where it is needed via the tracheoles directly to the tissues. This is known as a dry system, although at the tracheoles endings near the cell they are filled with fluid, and is many thousands times more efficient, allowing for the huge amount of oxygen required by the flight muscles to be available.

Dr Bailey, the famous British bee researcher, discovered that when in flight bees take in air from the first spiracle, which is the largest, and expel CO_2 from the third spiracle, both of which are on the thorax.

The air sacs are expanded and contracted by fluid pressure changes due to the movements of the plates of the abdomen caused by the muscles contracting and expanding; this allows for fresh air and stale air to be replaced. There is a 50% increase in oxygen consumption when bees fly from rest.

The sectional drawing shows the internal mechanism that controls the opening and closing of the valve by the muscle, which contracts to close the valve by pressing against the cuticular ridge. The image shows the valve in its open state.

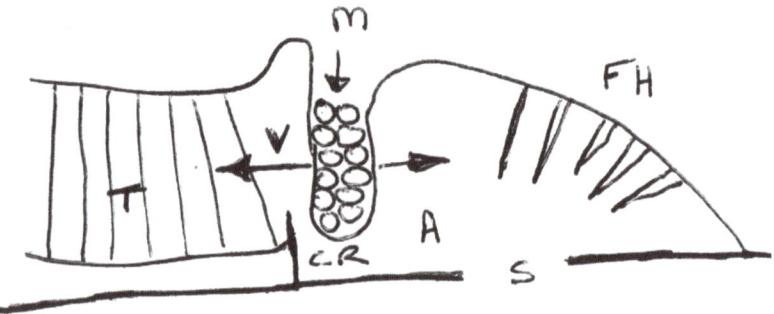

Fig 98. T trachea. V valve. M muscle. CR cuticular ridge. A atrium. FH filter hairs. S spiracle. The picture was drawn from Snodgrass.

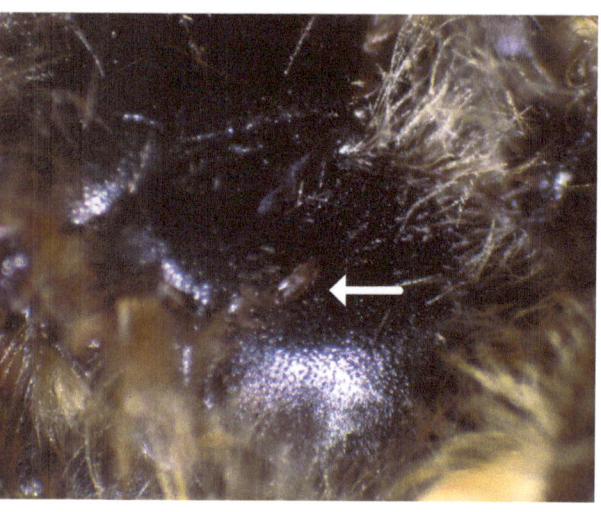

Fig 99. Thorax third Spiracle. X 40 magnification about 0.3 mm long.

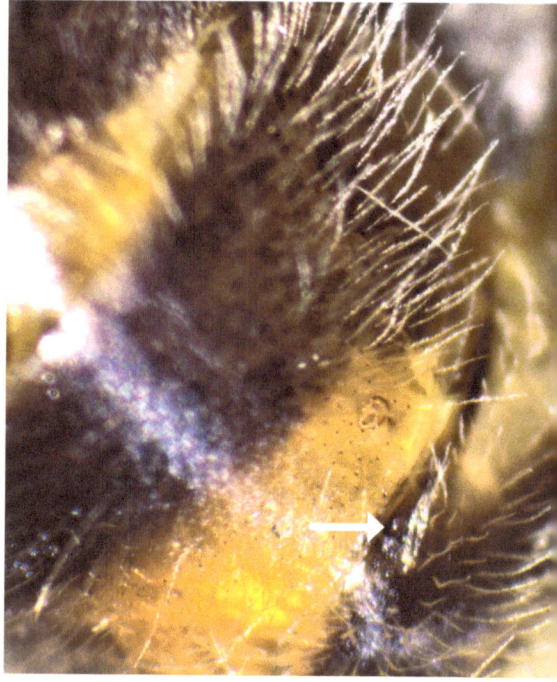

Fig 100. Abdominal 4th Spiracle. X 20 magnification 0.3 mm long.

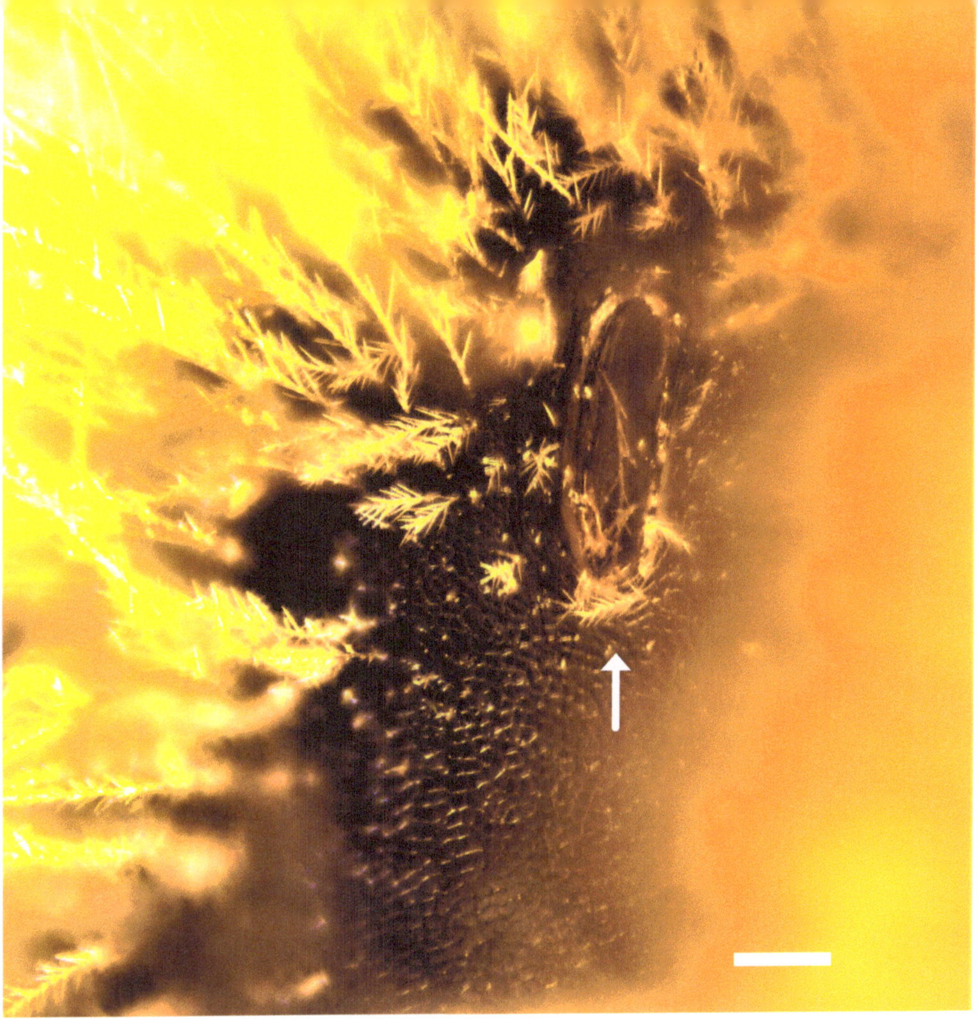

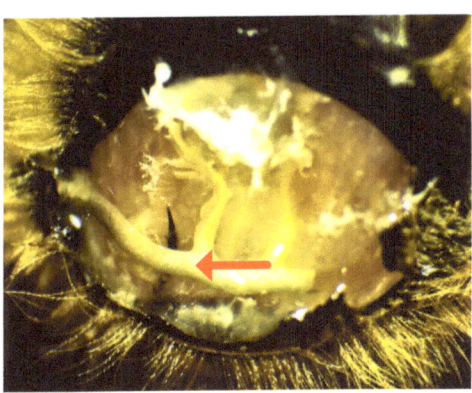
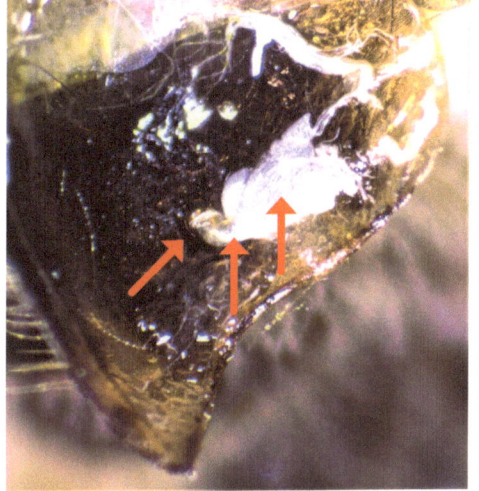

Fig 101 (top). Spiracle, scale bar 100 microns. X 200 Magnification.
Fig 102 (left). First Trachea on the thorax. No evidence of mite infestation. X 40 magnification.
Fig 103 (right). Internal view, spiracle, trachea and air sac. Third Spiracle on the thorax. X 40 magnification.

Drawing breath

The advantage of a constant direct airflow to cells, for example, when it flies, is that the high demand for oxygen by the flight muscles can be met. A wet system would be too slow to process the amount of oxygen required.

The air enters the ten pairs of spiracles on the outside of its thorax and abdomen, entering via a chamber - the atrium - which has some filter hairs inside, a muscle closes the chamber except for the first spiracle on the thorax which supplies the flight muscles. The third spiracle on the thorax is the largest.

Some of the air goes directly to the organs; however, there are many air sacs of various sizes found throughout their bodies, acting as reservoirs and buoyancy when flying. They take up a large surface area on the top of the abdominal organs and vary in shape and location in different bees. The brain has air sacs surrounding it.

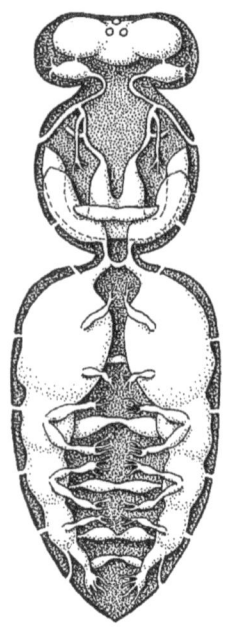

Fig 104. Tracheal sacs in the adult trunk. Taken from Dade by kind permission of IBRA.

The tracheae coming from the air sacs act like motorways conveying the air forward to smaller tracheae, then even smaller ones, called tracheoles, which meet with the tissues where respiration takes place. To stop the trachea from collapsing under pressure they have supporting soft, flexible cuticle bands wrapped around them, which are spaced out, providing rigidity. These are called Taenidia, meaning a ribbon. The function of respiration is to allow gaseous exchange. Oxygen (O_2) is provided to the cells to release energy. Carbon dioxide (CO_2), waste products of energy and water, are exchanged. This is done partly by diffusion at the end of the tracheoles, which are permeable, and mainly by the muscular action of the abdomen caused by the sternites and tergites of the exoskeleton that are shortened and lengthened. This drives the inhalation and exhalation of gases when energy is needed at a high rate. Due to the minute size of the tracheoles, gases can't transfer by mechanical movement. However, as the end of the tracheoles are embedded into the tissue, the bees' blood, haemolymph, is also present. Here muscular pressure forces the blood up to the end of the tracheoles, allowing O_2 to be supplied by diffusion in solution at a faster rate.

When a bee is at rest, it mainly breathes through the spiracle on the thorax. When flying, the air is drawn in on the thorax and abdomen but exhalation is mainly through the spiracle on the propodeum.

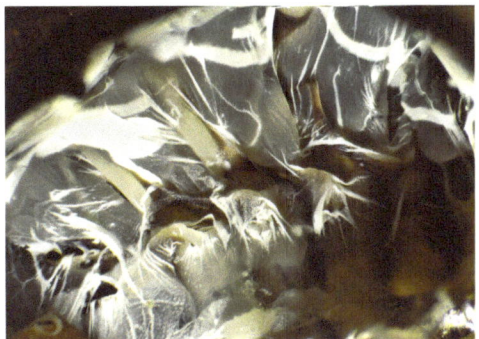

Fig 105. Air sacs abdomen.
X 40 Magnification.

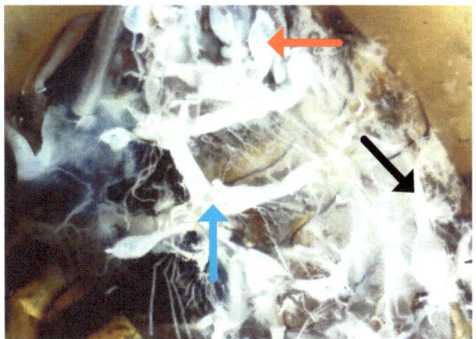

Fig 106. Worker bee tracheae, some air sacs removed. Air sacs, red arrow. Spiracle, black arrow. Trachea, blue arrow. X 20 Magnification.

Fig 107. Air sac in the abdomen, showing supporting network.
X 100 Magnification.

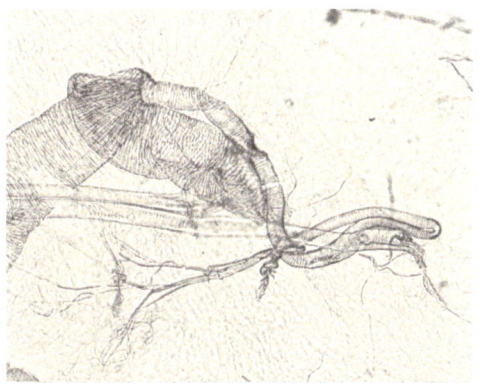

Fig 108. Tracheoles.
X 400 Magnification.

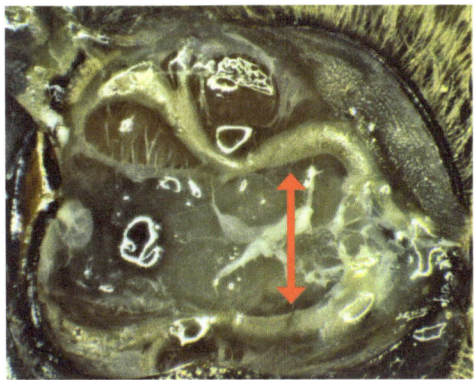

Fig 109. Main tracheae by spiracle one, this is where the Acarine mite will live and feed.

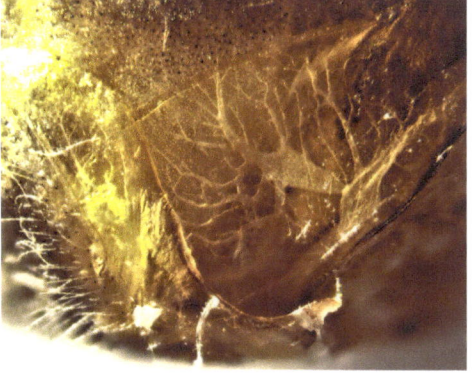

Fig 110. Tracheae from a live worker bee visible from the top of the abdomen, between the plates.

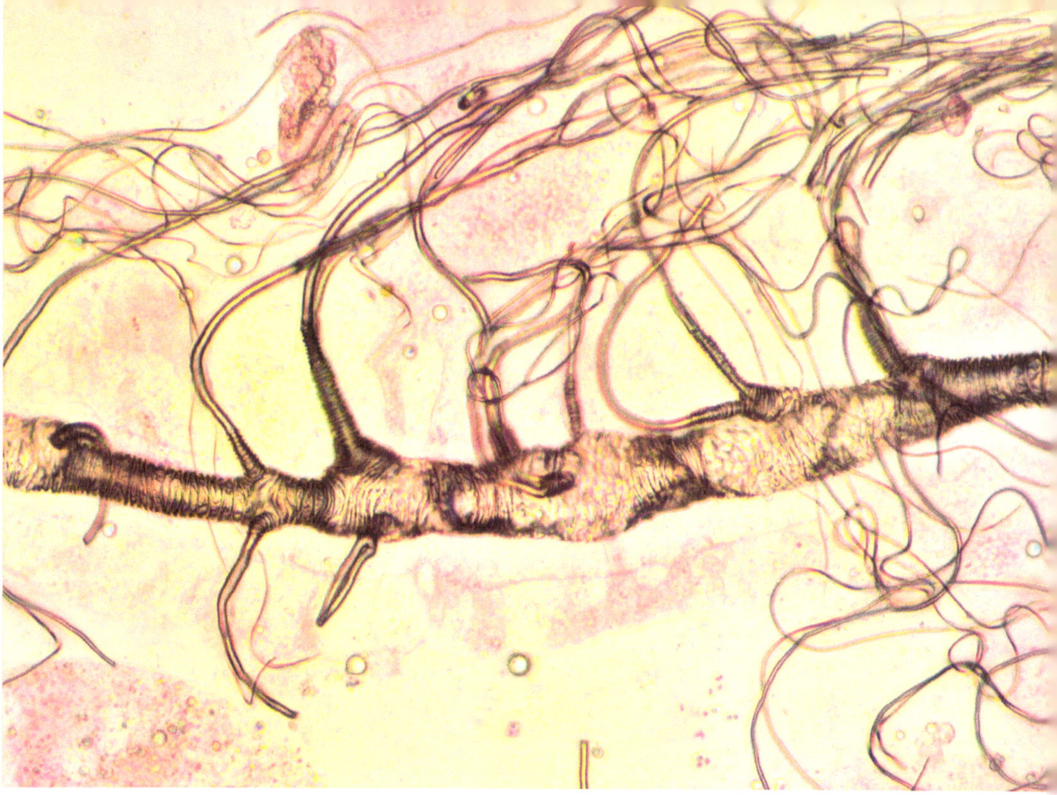

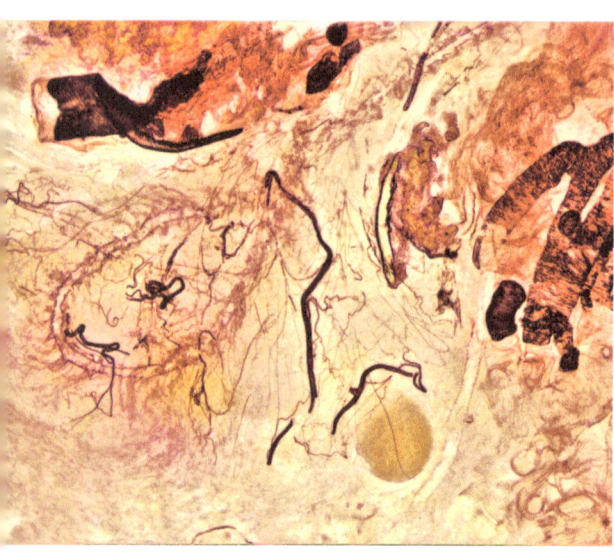
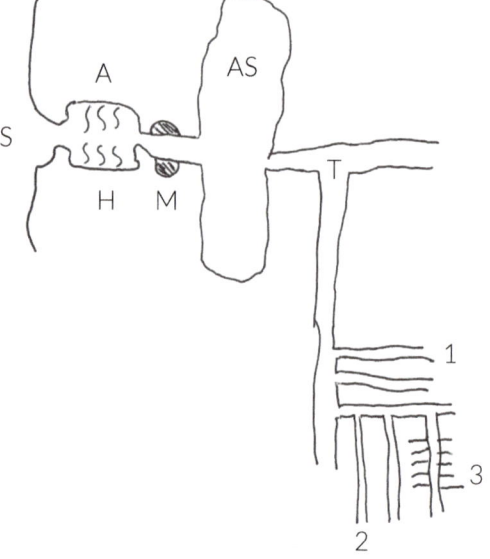

Fig 111 (top). Taken from a larva, showing supporting Taenidia.
Fig 112 (left). Queen bee tracheae and tracheoles were taken from spermatheca. Note the sizes of each tubular system displayed.
Fig 113 (right). Insect diagrammatic respiratory system. S Spiracle. A Atrium. H Hairs. T Trachea. 1. Small trachea. 2. Minute trachea. 3. Tracheoles.

The Ventral Nervous System

Ventral (adjective.)

1739, from French *ventral* or directly from Late Latin *ventralis* "of or about the belly or stomach," from Latin *venter* (genitive *ventris*) "belly, paunch; stomach, appetite; womb, unborn child."

Nervous (adjective.)

Late 14c., "containing nerves; affecting the sinews" (the latter sense now obsolete); from Latin *nervosus* "sinewy, vigorous," from *nervus* "sinew, nerve" (see nerve). The meaning "of or belonging to the nerves" in the modern anatomical sense is from the 1660s.

Nerve (verb.)

C. 1500, "to ornament with threads;" Meaning, "to give strength or vigour" is from 1749. Related: *Nerved*; *nerving*.

System (noun.)

1610, The whole creation, the universe," from Late Latin *systema* "an arrangement, system," from Greek *systema* "organized whole, a whole compounded of parts," from the stem of *synistanai* "to place together, organize, form in order.

The control centres, the ganglia

The nervous system develops in the larva and is retained through the development of the bee.

There are eleven ganglions (Greek for a tumour or a swelling) in the larva and an additional sub oesophagus ganglion that is found in front of the brain; it controls the mouthparts in the larva. Some ganglia merge during development at the pupal stage, e.g. four become fused together, resulting in two ganglia in the thorax. Two others fuse in the abdomen and this then makes a total of five ganglia in the abdomen.

The honeybee has a brain and dual nerve trunks running down its ventral side. At various points, there are seven ganglia.

Ganglia are nerve centres, made up of neurons - nerve cells that allow sensory and motor nerve cells to interact.

In the adult bee, sensory cells are found close to the epidermis and receive messages from the hairs, pegs and pits that make contact with the outer surface of the cuticle. Motor cells lie in the central nervous system with axon - branches to muscles.

Ganglia have some autonomous functions controlling their delegated areas such as the legs and wings. If

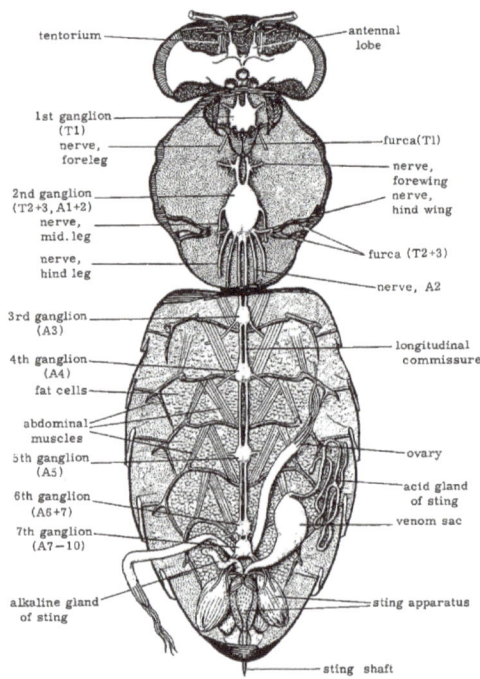

Fig 114. Worker nervous system, taken from Dade with kind permission of IBRA.

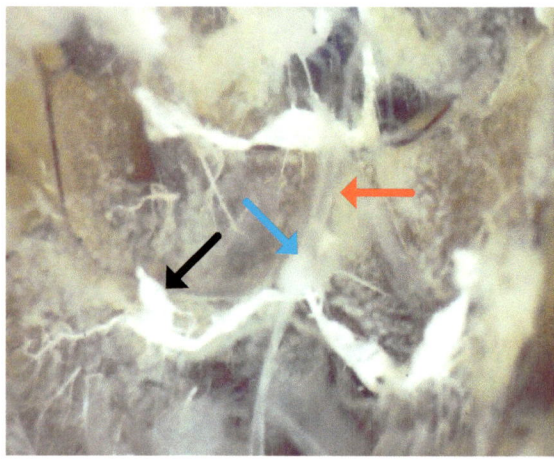

Fig 115. Internal view of the abdomen, viscera removed. Dual nerve trunk, red arrow. Ganglia, blue arrow. Trachea sac filled with air. Black arrow.

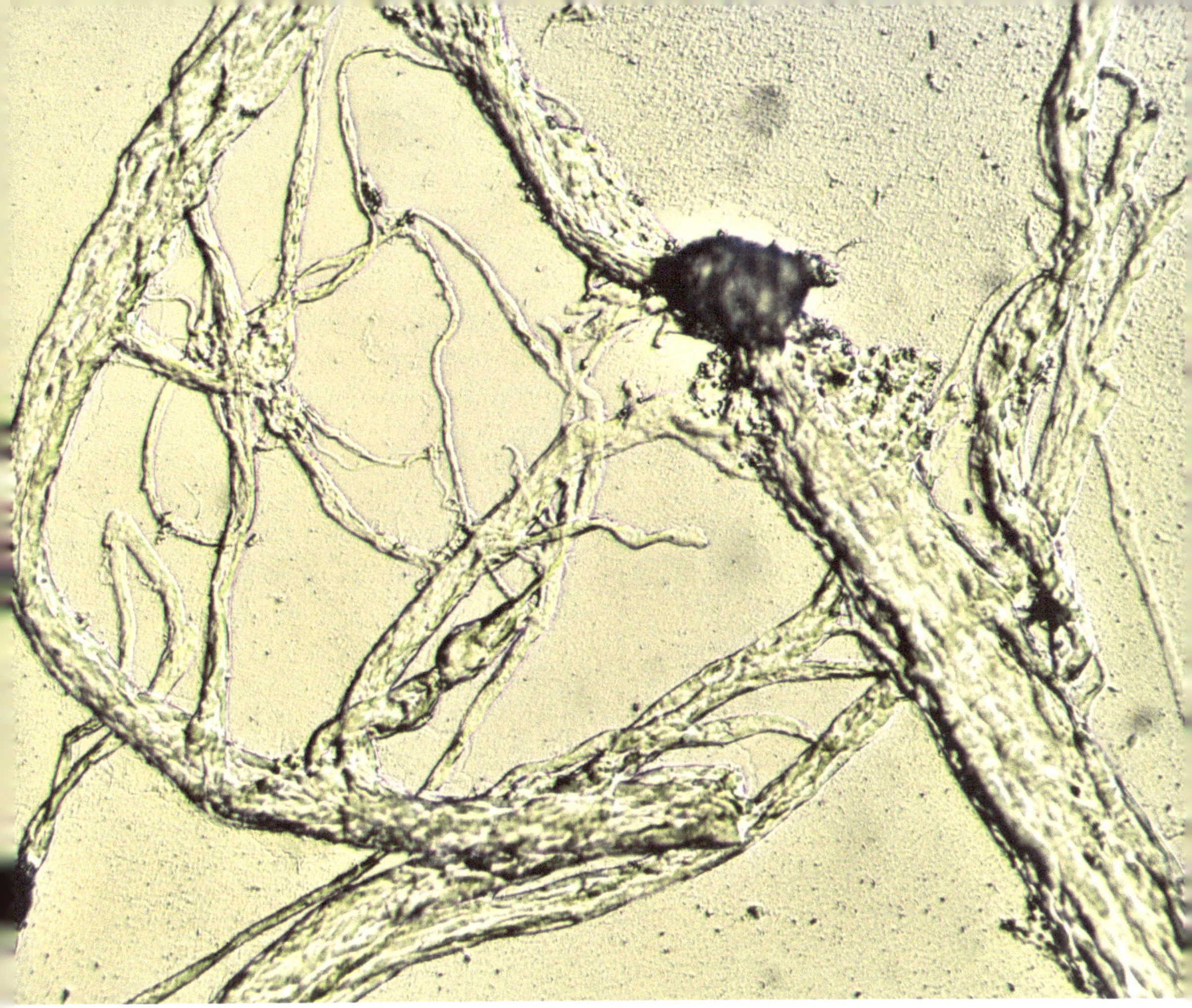

Fig 116 (above). Nerve fibres.
Fig 117 (right). Close up view of ganglion showing two ventral nerve cords and branching nerve fibres. Black arrows.

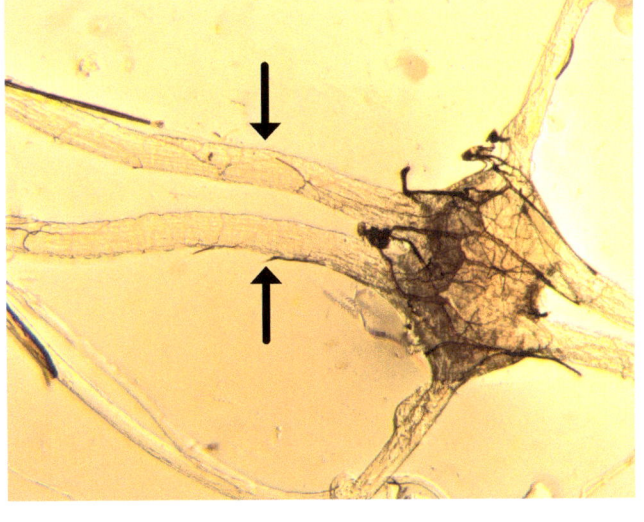

a bee loses its head it can still walk, move its wings and sting with the aid of these nerve bundles. However, it cannot coordinate its movements as the brain has overall control. The last pair of the ganglion is situated near the sting chamber and is ripped out when the worker stings; it is this ganglion that sends messages via the connecting nerves which cause the muscles to keep contracting, injecting more venom into its victim.

Each ganglion consists of two parts that have fused; they send out nerves to the other bodily areas throughout the body, including the sense organs. They are connected by a pair of twinned nerve cords, throughout the body called the longitudinal commissures.

The flight muscles of the bee are not controlled by the ganglia due to the speed of the wingbeats (around 180 beats per second); this is a too fast response for the motor nerve impulses of the nervous system.

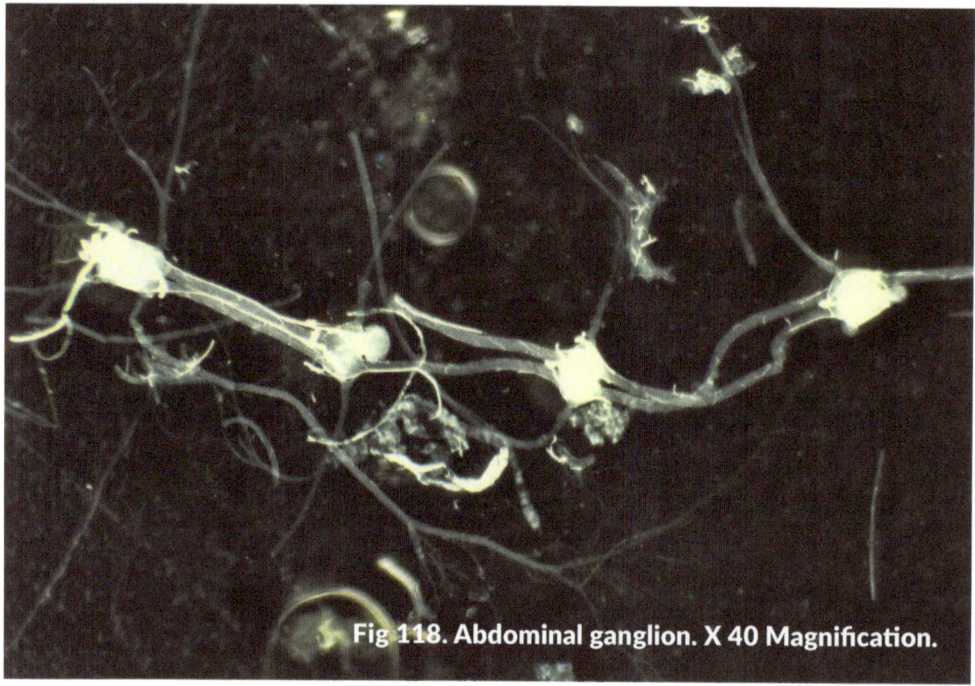

Fig 118. Abdominal ganglion. X 40 Magnification.

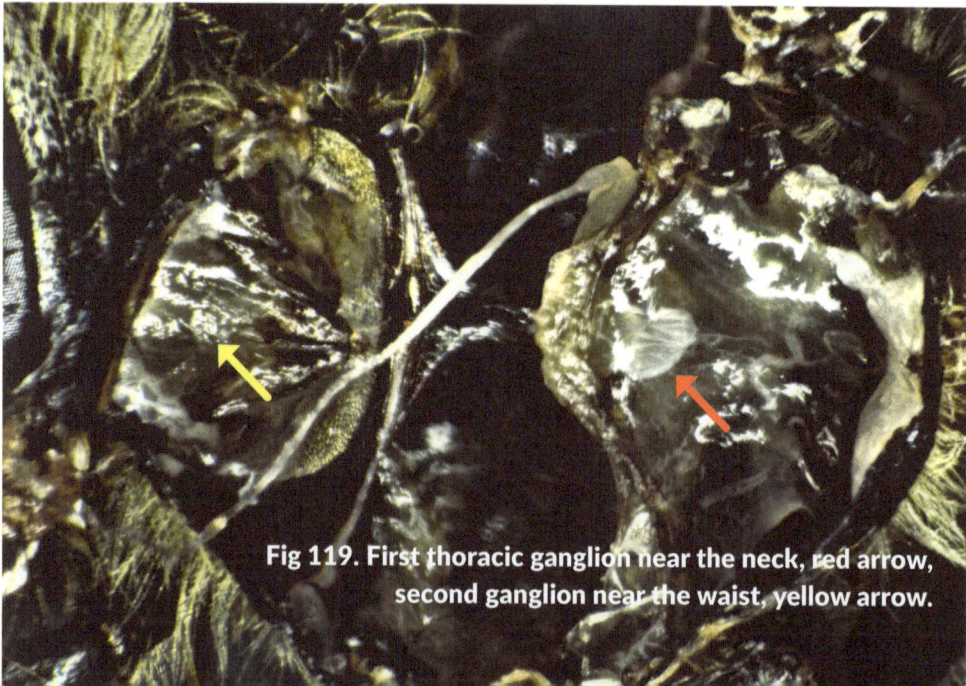

Fig 119. First thoracic ganglion near the neck, red arrow, second ganglion near the waist, yellow arrow.

Ganglion	Location	Area served
Supra-oesophageal = Brain	Head	Ocelli and compound eyes, antennae, labrum and Cibarium
Sub-oesophageal = Brain	Head	Mandibles and proboscis
G1	T1/T2	Forelegs
G2	Thorax 3/ Abdomen 1	Mid and rear legs, wings and segment A2
G3	A2	Whole of A3
G4	A3	Whole of A4
G5	A4	Whole of A5
G6	A5	Whole of A6
G7	A6	Whole of A7 and sting chamber

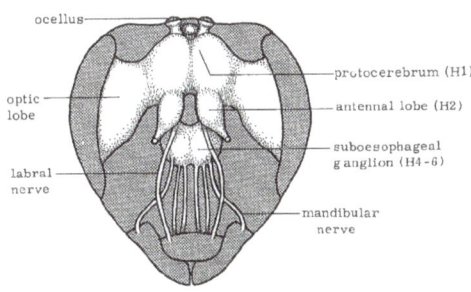

Fig 120 (above). The brain and principle nerves of the head, anterior view. Taken from Dade by kind permission of IBRA.

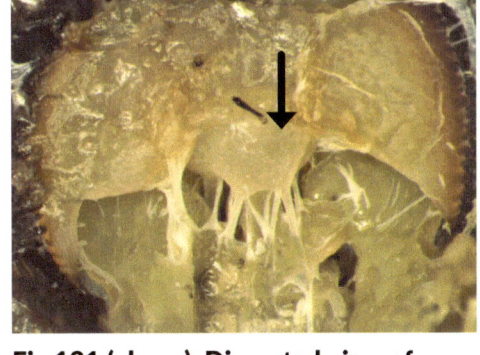

Fig 121 (above). Dissected view of the rear of the head of a worker bee, showing the suboesophageal ganglion. X 10 magnification.

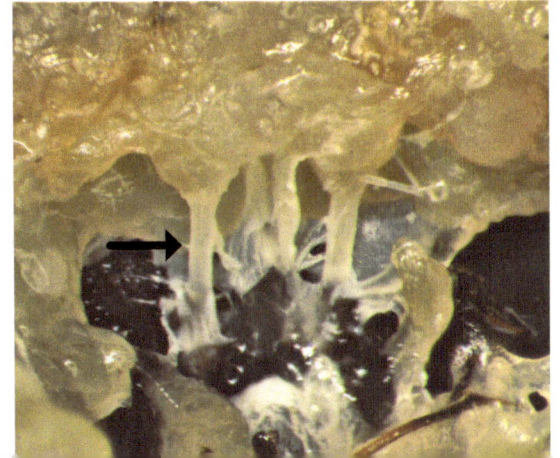

Fig 122 (left). A dissected view from the front of the head of a worker bee, showing the antennae lobes with descending antennae nerves. X 10 magnification.

The brain is largest in the worker bee; however, the optic lobe of the drone is much larger than that of the workers due to the increased eye size, which makes their brain appear bigger. The brain consists of three parts, the protocerebrum (first brain), deutocerebrum (second brain) and tritocerebrum (third brain). The major part of the protocerebrum is made up of the optic lobes which are crisscrossing nerve fibres connecting to the compound eyes. The deutocerebrum consists of bundles of nerve fibres connected to the antennae. The tritocerebrum is the smallest part and is hidden by the other parts; it sends nerves to the labrum and frons, the lips and forehead.

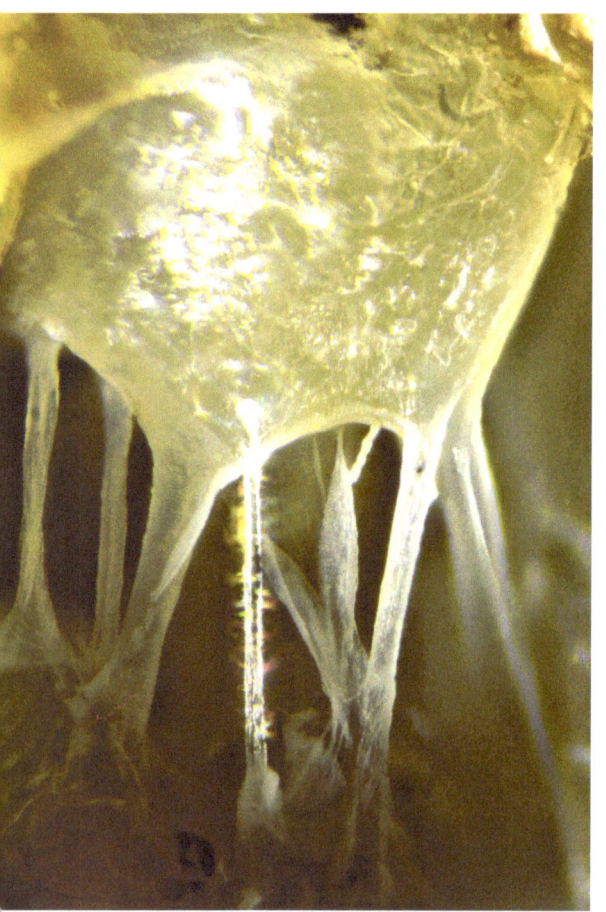

Fig 123. The suboesophageal ganglion, viewed from the rear of the head, showing the following descending nerves: labial, labral, mandible and maxillary. X 50 magnification.

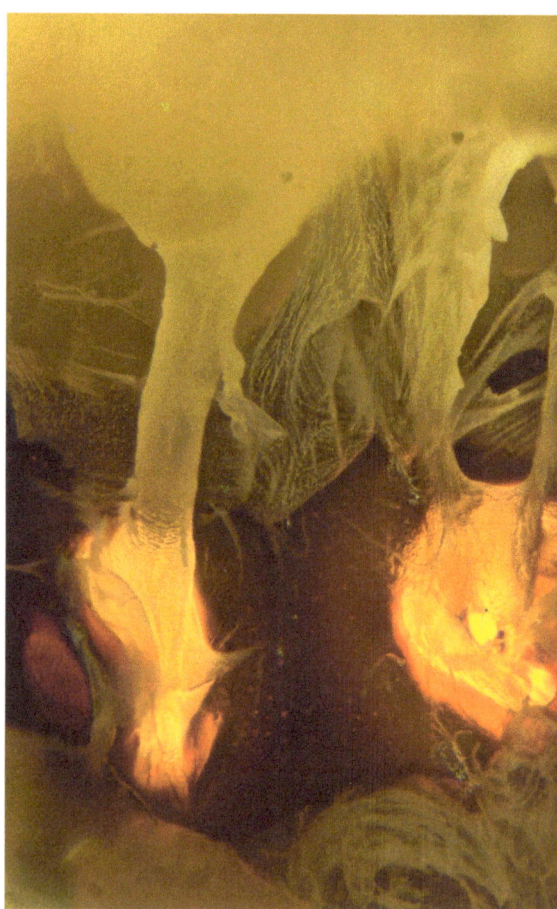

Fig 124. The antennae lobe showing antennae nerves. The reflected orange light from below highlighting the internal scape of the antennae (the ball joint). X 50 magnification.

Circulation

Circulation (noun.)

Mid-15c. *Circulacioun*, in alchemy, "process of changing something from one element into another," from Latin *circulationem* (nominative *circulatio*), noun of action from the past participle stem of *circulare* "to form a circle," from *circulus* "small ring".

Of blood, "act of moving so that it returns and begins again," first by William Harvey, 1620. Meaning "act or state of being distributed" is from the 1680s; that of "the extent to which a thing circulates" (of periodical publications) is from 1847.

Heart of the matter

The circulatory system of the honeybee, unlike mammals, is an open system and not a closed one and it does not need to pump oxygenated blood, via blood cells around its body as this requirement is met by the network of branching trachea that infiltrates close to the cells. This dry system meets their requirement for rapid oxygen transfer to aid in supplying the flight muscles. A closed wet system would never meet the high oxygen needs of the muscle cells.

The bee's blood, haemolymph, meaning blood and water in Latin and Greek, is the fluid that bathes the internal organs and plays a vital role in excretion, food distribution and hormonal messages.

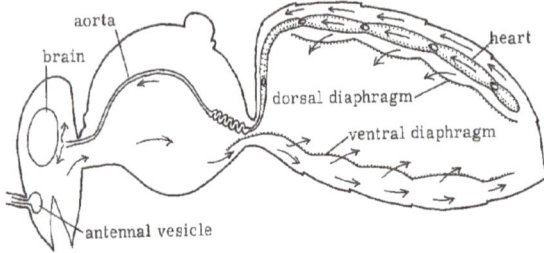
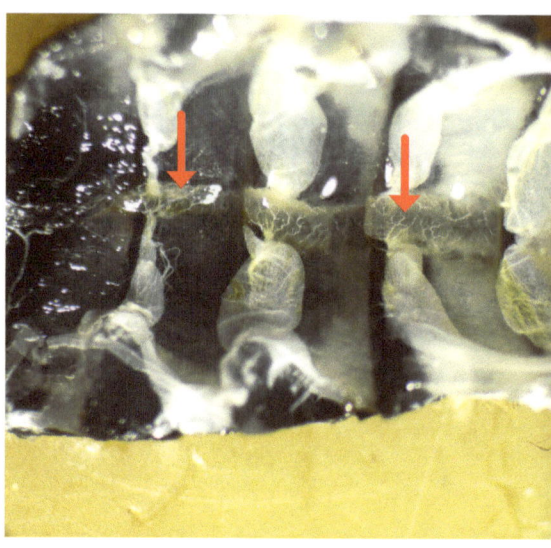

Fig 125. Circulatory system, taken from Dade with kind permission from IBRA.
Fig 126. Drone's heart red arrows, diaphragm removed, surrounded by air sacs X 10 Magnification.

The heart is a long muscular tube, which beats; this can be seen with the naked eye when dissecting an anesthetised bee. Scientists have developed a set protocol for anaesthetising the bee first. The heart is made of different parts that run on the dorsal side of the abdomen, surrounded by muscle cells that are capable of contracting in a forward rhythmic motion. It starts with a blind end and has five pairs of opposing oval valves, which have slit openings, named Ostia - meaning door - which when relaxed, allow the haemolymph to infiltrate. When the internal pressure rises in the abdomen, the heart Ostia valves close and prevent the haemolymph from returning to the abdomen allowing the blood to progress towards the head via the thorax. The heart is suspended from the dorsal (upper top side of the exoskeleton) attached by the branching threads of the diaphragm muscles.

The tube passes through the tiny waist called the petiole - a stalk - where there are nine loops, known as aorta loops. Scientists think they act as heat sinks - the returning blood from the thorax is warmer than the blood in the aorta, due to the large flight muscle activity, so warms the incoming blood to the head. The thorax and head are at a higher temperature normally due to flight muscle requirements and the abdomen is often closer to the ambient temperature. The aorta then continues through the centre of the thorax and the neck to behind the brain where it terminates in an open end.

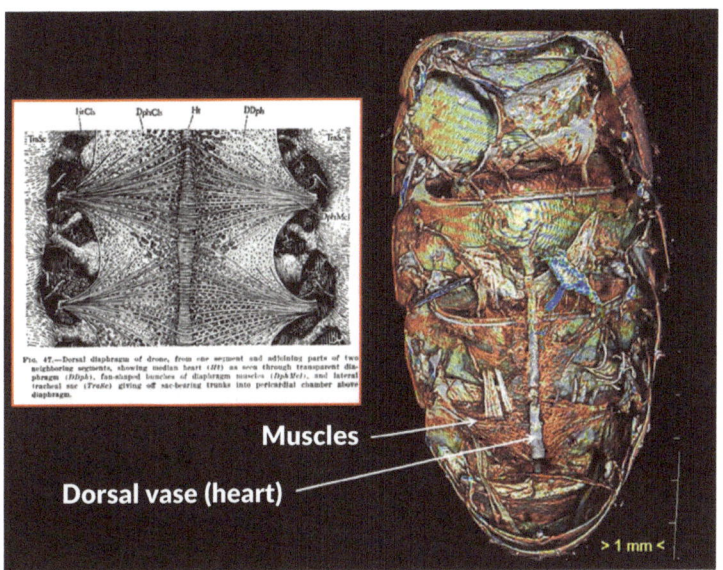
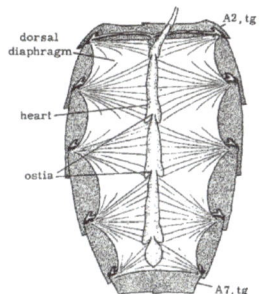

Fig 127. Heart viewed from underneath, showing openings, Ostia. Taken from Dade by kind permission of IBRA.

Fig 128. An amazing CT scan of honeybee showing heat and muscles as per Snodgrass by kind permission ©Prof. Dr Javier Alba-Tercedor. Univ. of Granada. Spain. Please visit this site for more fantastic images of the bee's anatomy. https://analyticalscience.wiley.com/do/10.1002/micro.2789/full

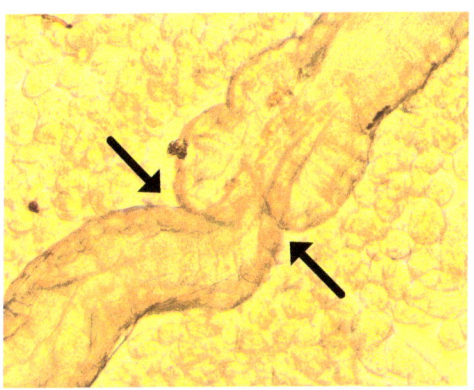
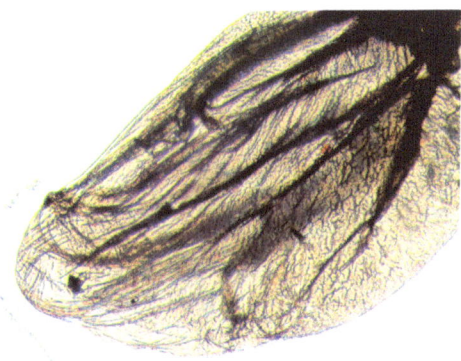

Fig 129. Close up of Ostia. X 200 Magnification.
Fig 130. Beginning of the heart blind end. Showing supporting tracheoles X 400 Magnification.

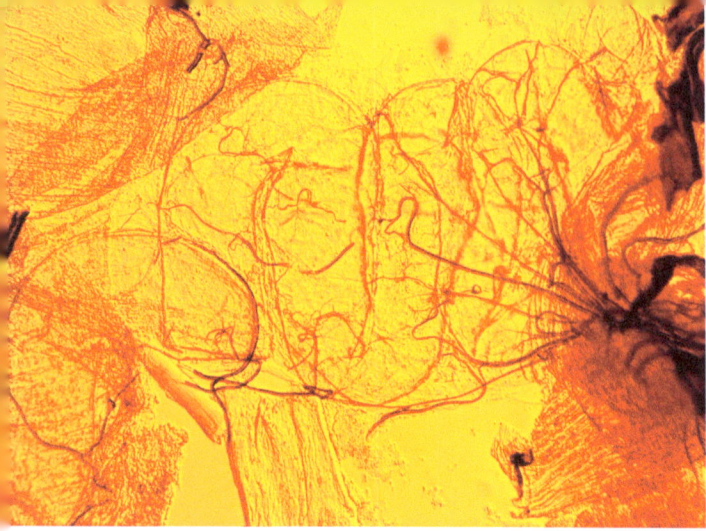

Fig 131. Loops of the aorta from a slide. X 200 Magnification.

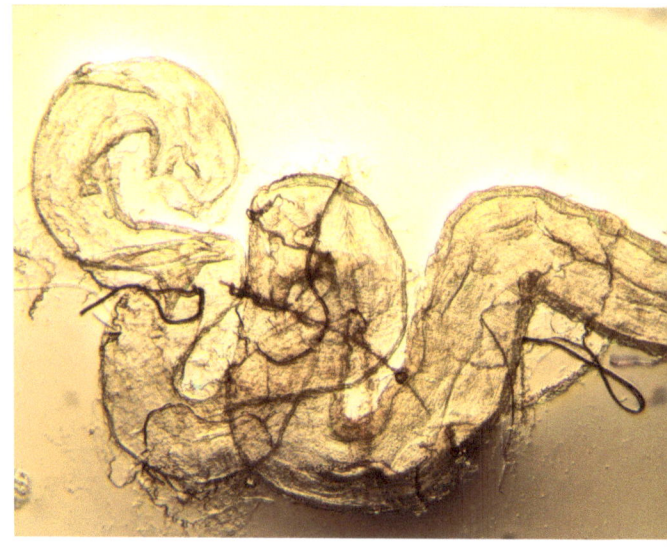

Fig 132. Loops of aorta dissected out. X 200 Magnification.

The haemolymph receives hormones from the neurosecretory cells of the brain and endocrine glands and distributes these throughout the body. The haemolymph also receives any waste metabolic compound from cells and transports them to the Malpighian tubules to be excreted. The wings and antennae are inflated by blood pressure and help to provide rigidity to their structures.

The returning haemolymph is free to circulate and find its way back to the abdomen where two thin diaphragms, one underneath the heart and the second at the bottom of the abdomen, help move the blood back towards the roof of the abdomen. The panting action when breathing aids in circulation. It is in the abdomen where the Malpighian tubes remove waste nitrogen from the blood and when it reaches the rectum, the nutrients and water are reabsorbed by the rectal pads back into the haemolymph.

A special pulsating vesicle, attached near the antenna in the head, provides pressure to pump haemolymph into each antenna for cellular needs. There are no muscles to support the antennae at the ends, so a specialised pump must supply the pressure to keep them turgid. This is named the antennal vesicle and consists of an ampulla, a hollow vessel with two tubes that run up each antenna. The walls of the vesicle have small perforations to allow blood access. There are no muscles attached to the vesicle but the dilator muscles and the pharynx muscles used to suck nectar, squeeze the ampulla, pulsating haemolymph up to the top of the antennae.

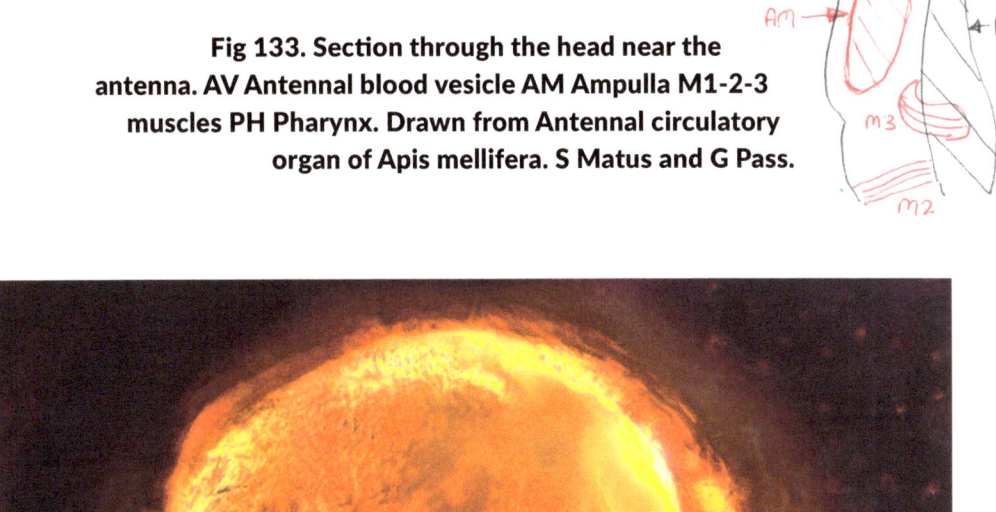

Fig 133. Section through the head near the antenna. AV Antennal blood vesicle AM Ampulla M1-2-3 muscles PH Pharynx. Drawn from Antennal circulatory organ of Apis mellifera. S Matus and G Pass.

Fig 134. Internal view of the antenna where it meets the head.

The legs are partly hollow and connected to the body cavity, so they naturally fill with blood; the leg movement helps circulate the blood within them.

Can bees suffer from a heart attack? The answer is no. They do not have blood vessels supplying the heart muscles with oxygen so they cannot suffer from the blocking of veins and arteries, the major cause of this disease in mammals. Furthermore, having an open circulatory system, there is no high blood pressure.

Exoskeleton

Exoskeleton (noun.)

In zoology, "any hardened external structure," as the shells of crustaceans or the scales and plates of fishes and reptiles, especially when it is of the nature of bone, 1841, from Exo + skeleton. Said to have been introduced by English anatomist Richard Owen.

How the bee got its hump

The basic shape of the thorax is rounded and is thought to be this strong shape to accommodate the massive flight muscles and the internal structures that support them. A wing is flexed up to 230 times a second when in flight. According to Kipling's 'Just So' story, the camel got its hump because of its laziness. Close observation of a bee's upper thorax at the abdominal end suggests it also has a hump, defined in the dictionary as 'A rounded protuberance on the back of a camel or other animal or as an abnormality on the back of a person.' You might have to move the hairs apart or look at an older bee to appreciate it fully. The comparison stops there, however! The bees are not lazy nor do they store food in their hump.

This protrudance is called the scutellum, derived from the Latin, scutal, little shield. It is a prominent ridge on the second segment, immediately behind the scutal fissure, a flexible joint that moves up and down when the flight muscles are active (see blue arrow in diagram). This allows the thorax to flex during wing movement - only the forewing is powered by the flight muscles; to which the hind wing is attached via the hamuli hooks.

The scutellum, which can vary in colour from yellow to dark brown and is one of the indicators when used to identify bee species, is used for air storage for use by the flight muscles as well as providing rigidity to the thorax and its internal muscle supporting structures, allowing the thorax to flex in flight.

Other insects have larger, more enveloped shields. Those from the Scutelleridae family or stinkbugs have an enlarged thoracic scutellum that forms a continuous shield to protect both abdomen and wings.

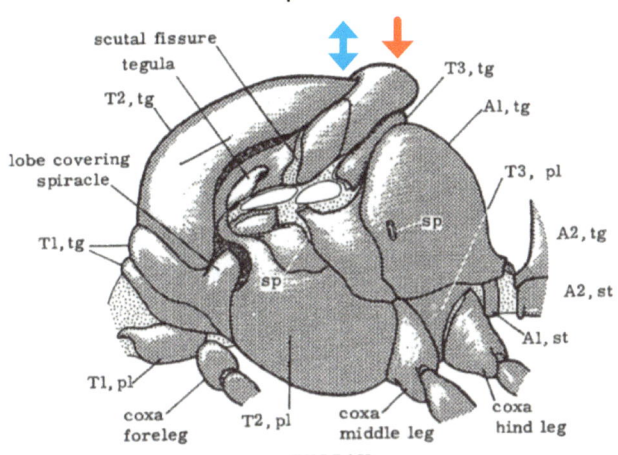

Therefore, like the camel, the bee does use its hump, but for air storage, albeit only for short periods of respiration.

Fig 135. Red arrow showing the scutellum, taken from Dade by kind permission of IBRA.

Fig 136A. Viewed from the top, shaved worker bee. X 10 Magnification.

Fig 136B. Viewed from the side, shaved worker bee. X 10 Magnification.

Fig 136C. Sections through thorax showing hump with air sacs and below the flight muscles. X 25 Magnification.

Fig 136D. Side view of the hump.

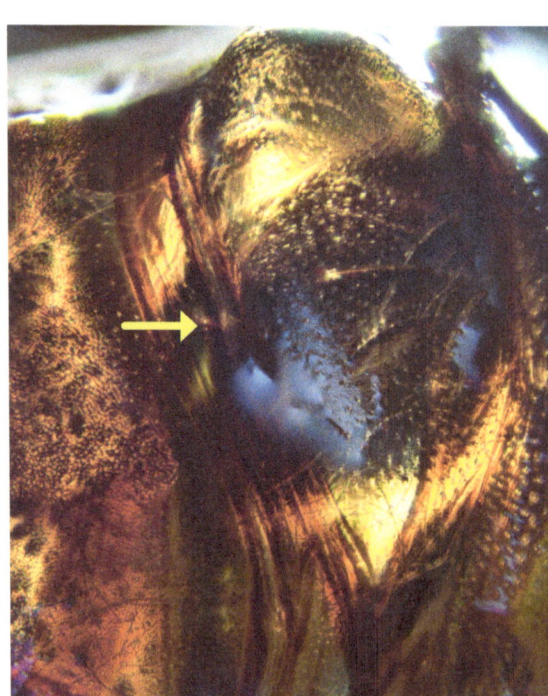

Fig 137. The scutal (shield) fissure, a flexible joint that moves up and down when the flight muscles are active.

An hourglass figure

Wasps, ants and bees have a narrow constriction joining their thorax to their abdomen named a petiole, meaning a stalk, formed between the first two segments of the actual abdomen; in the bees, wasps and ants, the first abdominal segment is fused to the thorax and is called the propodeum. Entomologists, when discussing the body of a bee in a technical sense, refer to the mesosoma (middle body) and metasoma (after body) or gaster (a term for the abdomen, taken from the Greek for the belly) rather than the thorax and abdomen, respectively. The petiole has a triangular shape when viewed from the top of the bee.

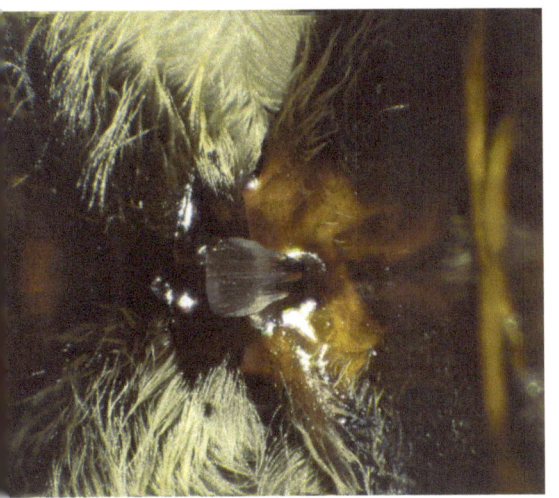

Fig 138A. Petiole of a worker bee, viewed from above. X 10 Magnification.

Fig 138B. Petiole of a drone bee, viewed from above, showing flexible membrane. X 40 Magnification.

The waist is narrow for flexibility. This area needs to act as a hinge, but unlike a human joint it is not surrounded by flexible skin; instead, it has a hard outer casing and a thin membrane that can flex sufficiently to enable the abdomens to curl and deposit eggs or sting. The drone's abdomen has a natural curve to enable mating on the wing. The drone is about two and a half times larger than a worker and has a larger petiole, which is easier to dissect.

The hairs adjacent to the petiole help the bee to know how far up and down the abdomen moves.

Fig 139. Worker bee showing part flexed abdomen.

The two tracheae pass through the petiole on either side of the top of the inverse triangle. (See diagram) Beneath them the oesophagus and aorta pass side by side, the ventral diaphragm sits beneath all four and at the point, the trunk nerves pass.

Also, two top muscles that levitate and turn the abdomen and their tendons pass through in the top of the waist and two depressor muscles pass through at the bottom.

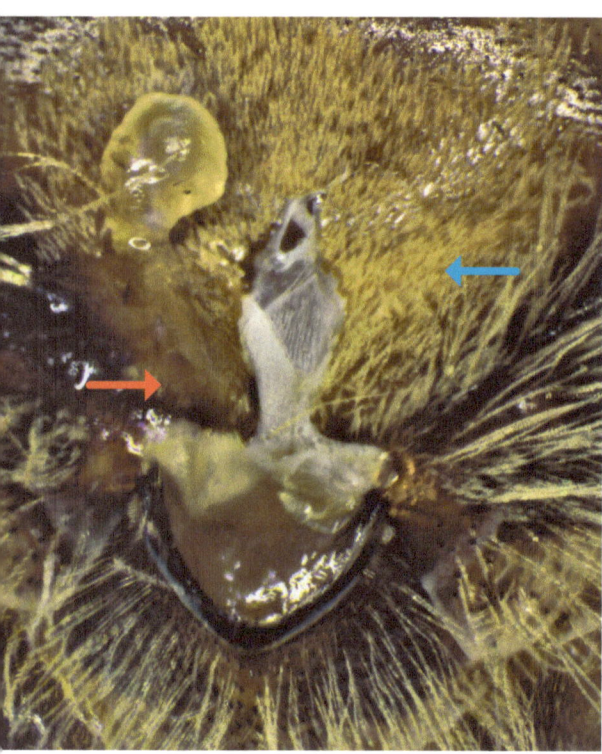

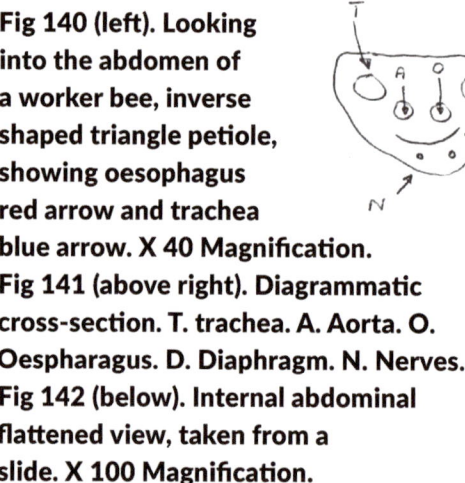

**Fig 140 (left). Looking into the abdomen of a worker bee, inverse shaped triangle petiole, showing oesophagus red arrow and trachea blue arrow. X 40 Magnification.
Fig 141 (above right). Diagrammatic cross-section. T. trachea. A. Aorta. O. Oespharagus. D. Diaphragm. N. Nerves.
Fig 142 (below). Internal abdominal flattened view, taken from a slide. X 100 Magnification.**

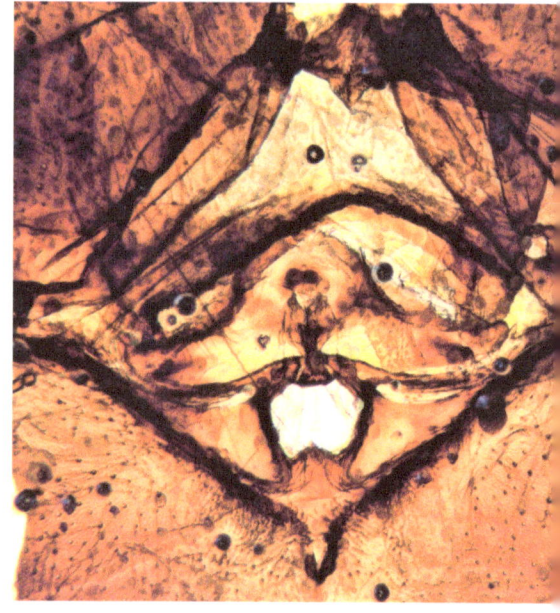

The thorax of a typical worker is 3.6mm wide and the petiole is 0.2mm wide, a ratio of 18:1. This would equate to a human with a chest size of 915 mm (36 inches) having a waist of just 51 mm (2 inches).

Hairy beasts

Bees have hairs, called plumose (due to their branching like structure) designed primarily to trap pollen. The average bee is covered in three million of these tiny hairs, even on its eyes, although the amount of hairs on the bee's legs is five times denser than on its eyes. The hairs are strategically placed to ensure the insect picks up as much pollen as possible. The gap between each hair is the width of a dandelion pollen grain (about 45 microns) which enables the bee to sweep up nearly 15,000 particles of pollen from one flower to the next, effectively enabling the insect to carry up to 30% of its body weight in pollen grains.

Even the drones have an abundance of hairs.

The state and age of a bee can be judged by its appearance. Nurse bees are normally fully covered in hairs, sick bee with viruses display deformed wings and a bald top to their thorax; often shiny, ageing worker bees have much fewer hairs on their bodies and tatty wings due to wear and tear.

There are other hair-like structures throughout the bee's body designed for specialist functions; they tend to be short peg-like, and for example, there are some on the top of the thorax on either side of the head. When the bee moves its head from side to side, they deflect and act as sensors to tell the bee how far it can rotate its head.

A honeybee uses her hairy front and middle legs like brushes to comb the pollen off her body and pack it into special hairy recesses on her rear legs called pollen baskets or corbiculae, which are specialized structures present only on the hind legs of the workers.

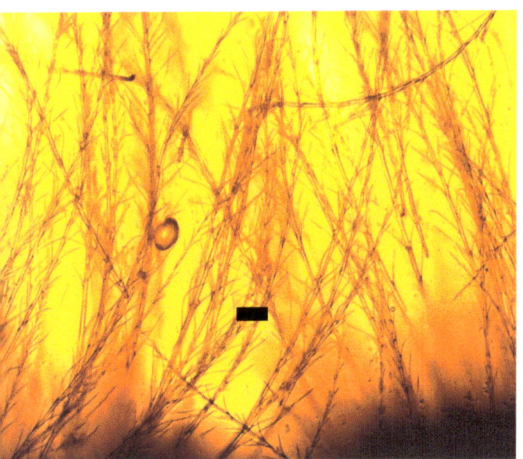 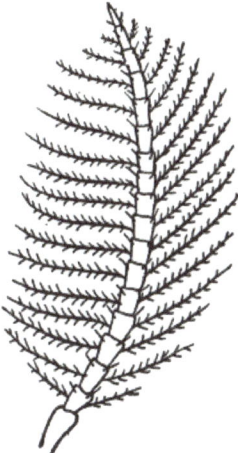

Fig 143A (far left). Plumose hairs showing pollen grain. Scale bar 25 microns. X 400 Magnification.
Fig 143B (left). The branch-like structure.

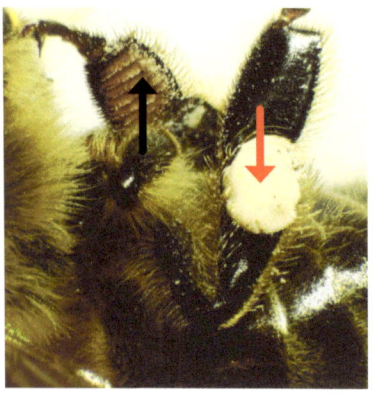
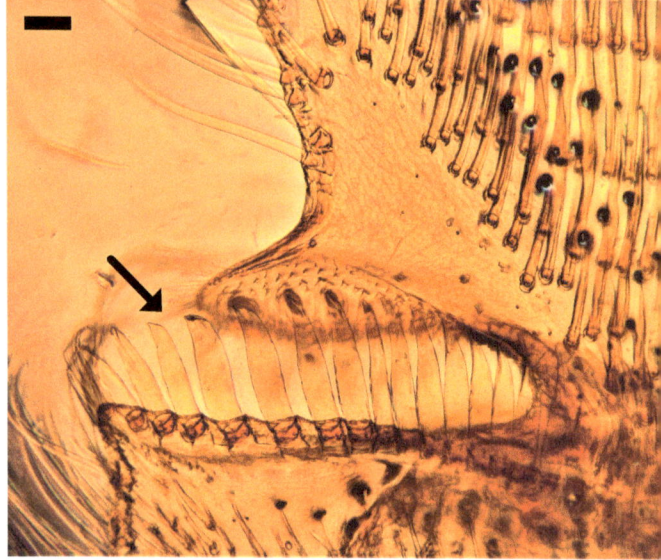
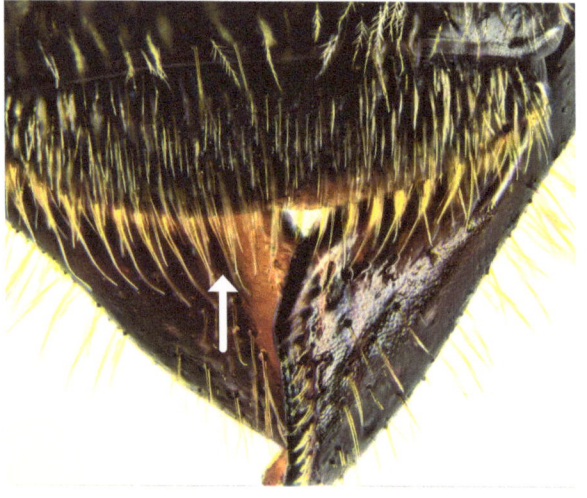

Clockwise from top left:
Fig 144. Healthy, hairy worker bee.
Fig 145. Deformed Wing Virus, showing polished thorax.
Fig 146. Cleared specimen showing specialised hairs on the hind leg. The rastellum. Scale bar 100 microns.
Fig 147. Specialist hairs on the top lip of the labrum, which help sense the movement of the mandibles.
Fig 148. Pollen basket, red arrow, black arrow, showing pollen brush on opposite leg.

Armour and Anchors

The honeybee has an external rigid frame, or exoskeleton, made up of hardened chitin, called the cuticle. This has no outward projections on its mainframe; however, during development it allows several different shapes to grow internally. These structures are referred to as apodemes, from the Latin meaning away from the body.

Shaped in the form of ridges, ribs, spurs and spines, the main role of the apodemes is to provide an anchor point for muscles and tendons. Some sclerotised regions of the external cuticle have grooves called sulci, (Latin for furrow) which are formed from internal ridges that strengthen the skeleton, a bit like internal roof beams. They anchor muscles and do not delineate any division of the skeleton.

The abdomen is divided into seven sections. The first, A1, is fused to the thorax. Known as the propodeum, (meaning 'in front of the stalk' or 'the waist') and it is often confused as being part of the thorax. The slim waist or petiole (Latin for a stalk) occurs on abdominal section two and is unique to hymenopterous insects. The abdominal sections two to seven form the rest of the abdomen, known as the gaster (Latin for the stomach).

The bee has from top to bottom three outer layers of plates, the upper plates of which are called tergites, the middle plates pleurites, while the lower plates have six sternites. Pleurites are absent on the bee's abdomen, though present on the thorax. When looking side-on, the tergites wrap around the top two-thirds of the abdomen and overlap each other. The tergites have colouration and hair growth patterns that vary across the various subspecies of the honeybee.

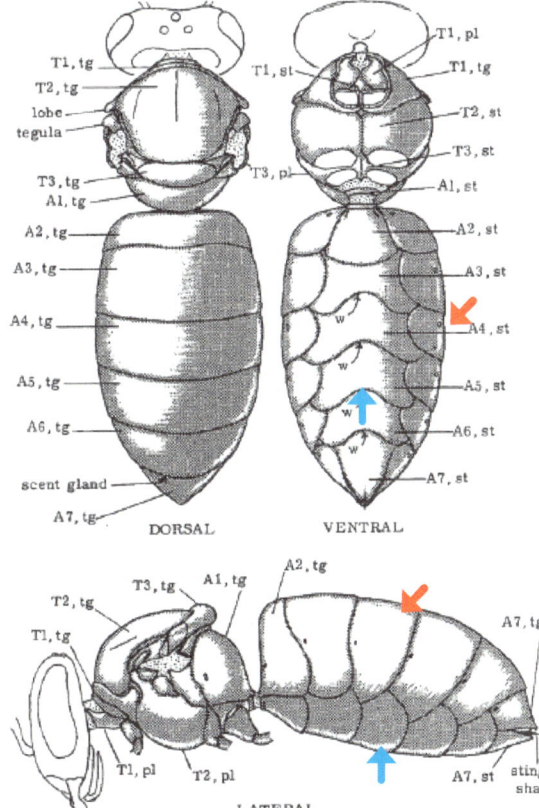

Fig 149. Thorax and abdomen numbered regions of the worker. A= abdomen. T= tergite, red arrows. Sternites, blue arrow. Taken from Dade by kind permission of IBRA.

Fig 150. Tergites of a worker bee.

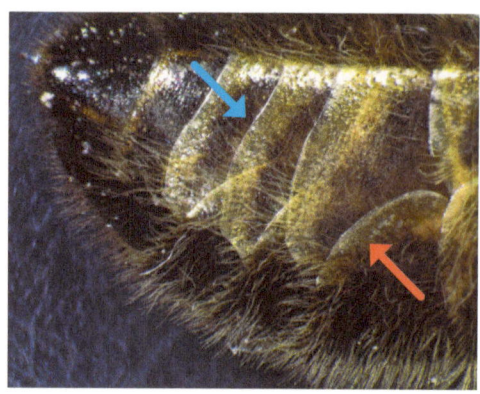

Fig 151. Tergites wrapping round, red arrow, and sternites blue arrow. The ventral side is the top view as seen.

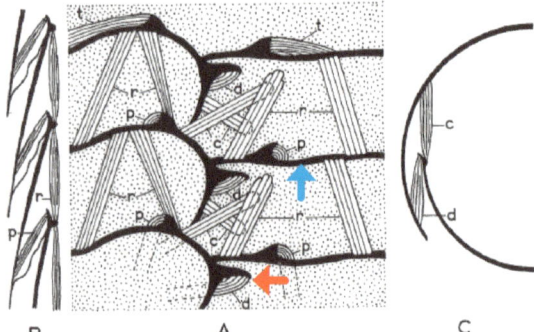

Fig 152. A section of the principle abdominal muscles taken from Dade. Red arrows showing hooks, blue arrow showing apodemes. Taken from Dade by kind permission of IBRA.

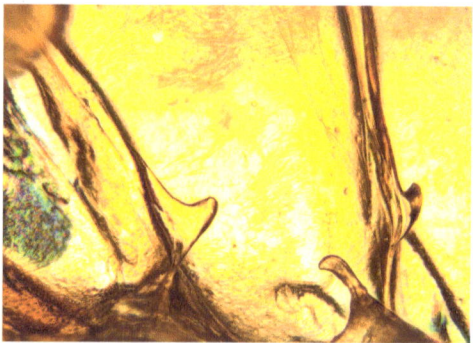

Fig 153. Internal abdominal hooks, these hooks are called apodemes; they form attachment sites for the musculature.

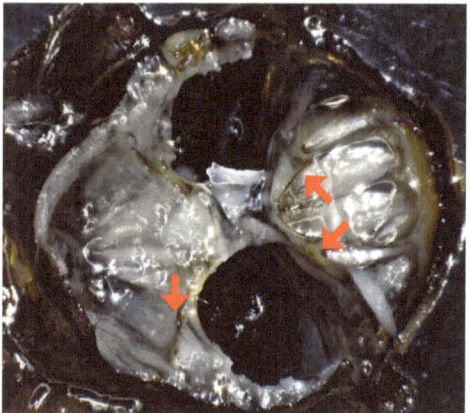

Fig 154. Preserved specimen showing an internal view of thorax showing supporting structures, the furca with muscles attached and a cleared view, red and yellow arrows.

Fig 155. Preserved bee. White arrow, endosternum. Yellow arrows, furca.

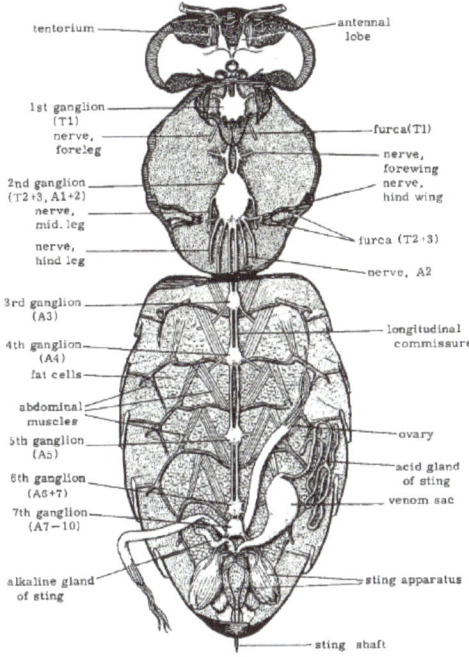

Fig 156. Showing floor of the abdomen and supporting structures, together with the thorax, with only the top muscles, removed. Taken from Dade with kind permission of IBRA.

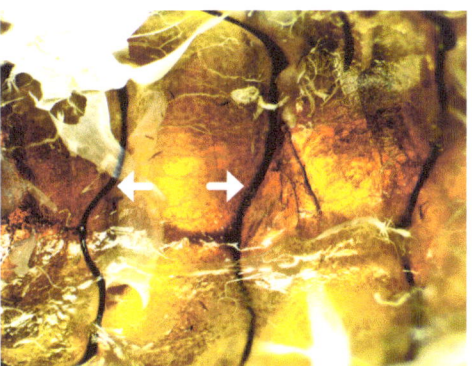

Fig 157. Apodemes inside of the abdomen of the worker.

Fig 158. Thorax of a drone showing the sclerotised regions of the external cuticle that has grooves in called sulci.

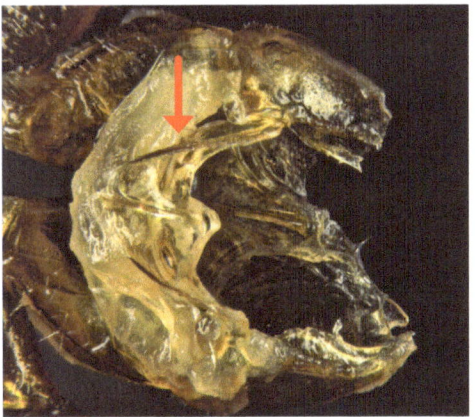

Fig 159. Cleared section of where the front legs join the thorax showing supporting structures.

Fig 160. Cleared section from beneath the head, showing supporting structure. Tentorial bridge, red arrow. Anterior Tentorial arm, black arrow. Tentorial is Latin for tent pole.

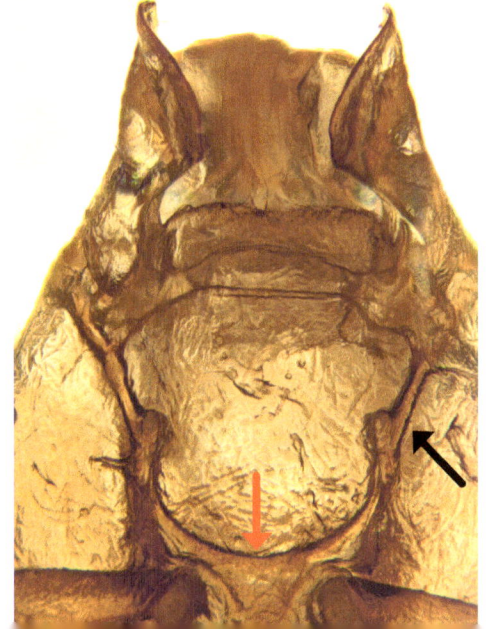

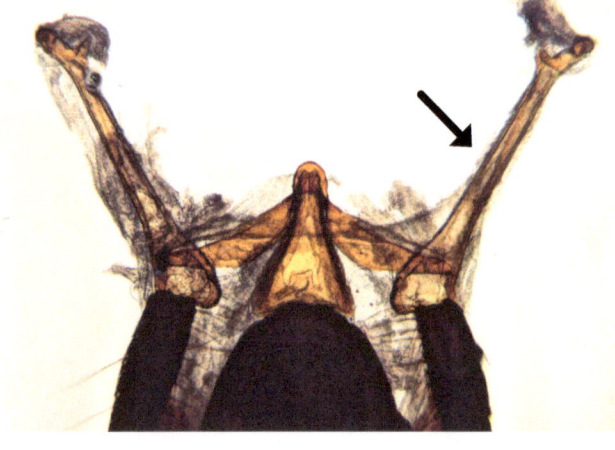

Fig 161. Dissected specimen showing the internal cardines, which act as levers, moving the proboscis backwards and forwards.

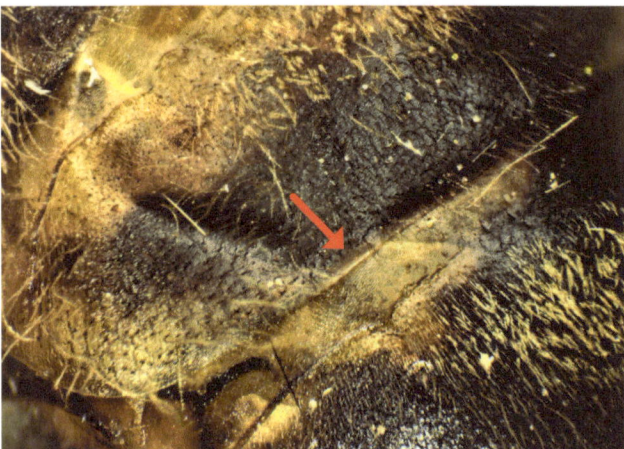

Fig 162. The joint between the tergites on the abdomen, showing the flexible intersegmental membrane.

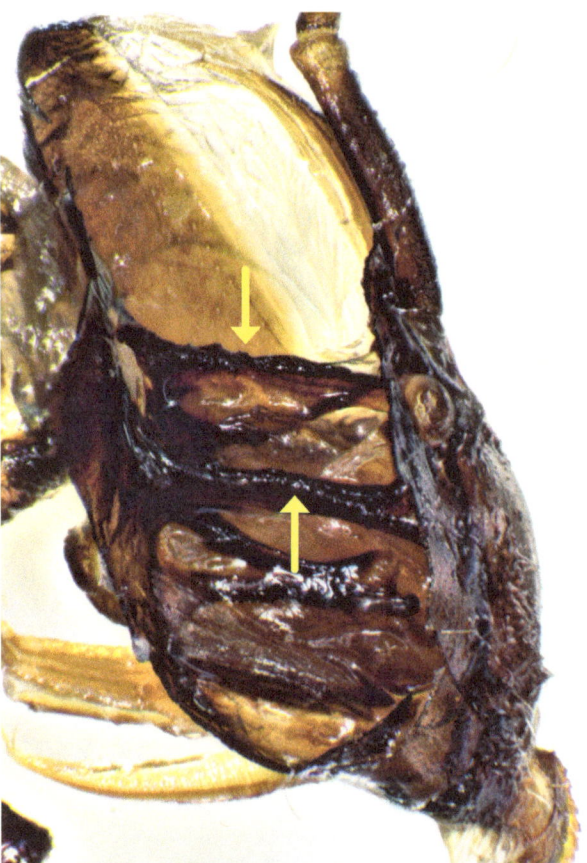

Fig 163. Cleared specimen-showing Tentoria inside the head.

Excretory system

Excrete (verb.)

"to throw out or eliminate," specifically "to eliminate from a body by a process of secretion and discharge," 1610, from Latin *excretus*, past participle of *excernere* "to sift out, discharge," from ex "out" + *cernere* "sift, separate". Related: *Excreted; excreting*.

Who was Malpighi?

Marcello Malpighi was born in Italy in 1628; he was primarily a doctor who, in later life, turned his attention to other sciences and is famed for many discoveries in medicine, botany and entomology. In 1669 in Britain, in recognition of his outstanding studies, he was honoured by being invited to join the Royal Society of London.

He was one of the first users of the microscope. It was whilst examining silkworms (the only domesticated insect) that he identified their tubules (small tubes) at the rear of the ventriculus (the stomach) as the organs that filtered liquid waste from the blood: similar in function to the mammalian kidneys. Insects share this common feature and his name has been given to them, Malpighian tubules.

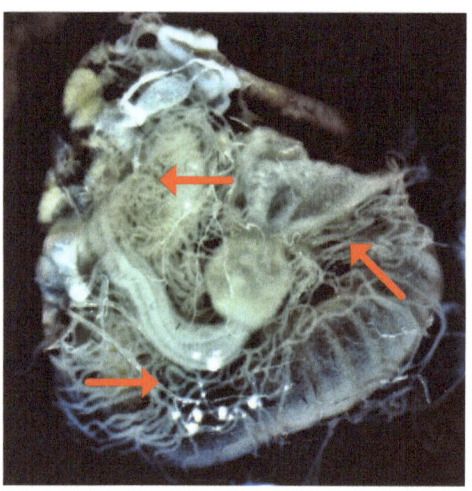
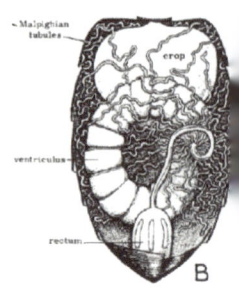

Fig 164 (left). Malpighian tubes of a worker bee convoluted, whitish, spaghetti-like structures. X 10 Magnification.
Fig 165 (right). Horizontal section of worker dorsal aspect. Taken from Dade with kind permission from IBRA.

About 100 tubules are floating in the abdomen; they open separately at the base of the ventriculus and the small intestinal junctions near the pyloric valve. (Sometimes referred to as the gatekeeper of the stomach) They vary in length and colour dependent on age and condition.

The bee's excretory system is essentially a sophisticated filtration system that not only removes waste substances that would slowly poison the cells but also acts selectively, adjusting the amounts of particular substances in the haemolymph so that there is a balance between water and salts. The osmotic pressure and acidity remain within narrow limits. There are two types of waste produced by active cells: carbon dioxide produced because of respiration, which is removed via the tracheal system and expelled through the spiracles, and nitrogenous waste resulting from the chemical reactions that go on within the cells involving proteins and other nitrogen-containing compounds. This in its basic form would be ammonia, which is toxic and requires a lot of water to remove it from the system. However, chemically present as uric acid, it is easily excreted and removed by the Malpighian tubules and then entered into the intestines. The first stage of filtration takes place when substances in the haemolymph are filtered through the wall of the Malpighian tubule at its upper (distal) end. Muscle fibres cause the tubule to float about and encountered the maximum amount of haemolymph throughout the abdomen. The substance passes passively (against its concentration gradient) or actively (by transport proteins in the cell membrane) across the single-cell layer of the Malpighian tubules and travels down the central cavity (lumen) of the tubule.

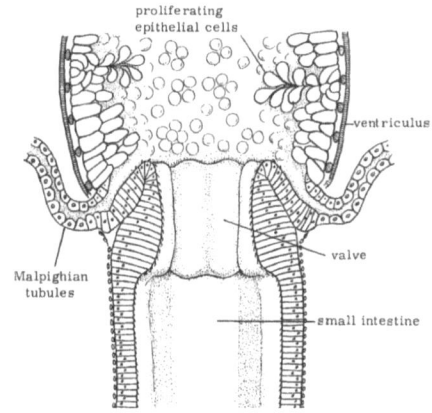

Fig 166. Section through the pyloric region, showing the junction between the proventriculus and the small intestines, the hindgut. Showing where the Malpighian tubules meet the valve. Taken from Dade by kind permission of IBRA.

The second stage takes place as materials travel down the lumen where some are re-absorbed; water retention is a crucial part of reabsorption - depending on the haemolymph state, salts may or may not be re-absorbed.

Finally, other substances are actively secreted by cells. The active transport of molecules from the haemolymph into the Malpighian tubules requires a lot of energy in the form of adenosine triphosphate, also known as ATP, which is a molecule that carries energy within cells, all living things use it hence the proximity of tracheoles for providing oxygen.

When the material in the lumen of the tubule reaches the intestinal connection it passes into the digestive system. Together with the waste from the digestive system, it is passed to the outside of the body through the small intestine, then rectum and finally the anus, where during its final passage further water is re-absorbed from the faeces via rectal pads on the anus.

Malpighamoeba mellificae, a disease of adult worker bees that affects the lumen of the Malpighian tubules, is a single-celled parasite, an amoeba that finally is encysted. It causes a contagious disease called amoebiasis, which can ultimately lead to the death of the host. It is commonly found together with Nosema infections. Due to its small, size of 3 - 15 μm it can only be

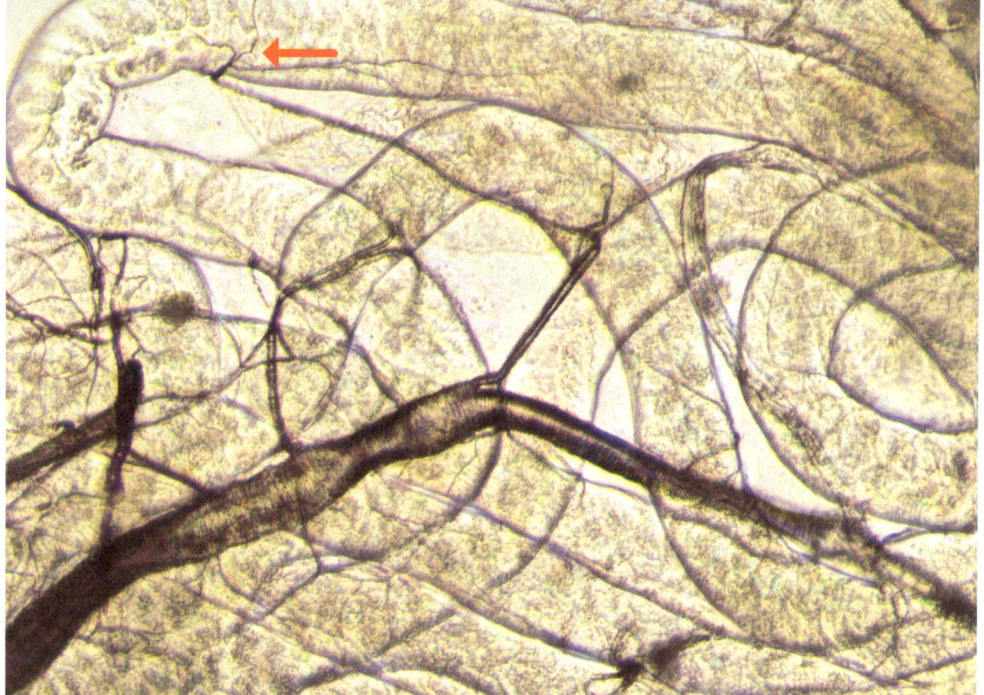

Fig 167. Malpighian tubules from the worker, showing supporting tracheal system, red arrow. X 200 Magnification.

seen and diagnosed by looking through a microscope at high magnification. Once encysted, the amoeba can pass down the tubules into the rectum to be ejected as faeces. The house bees then ingest the waste in the cleaning routine, reinfecting themselves. The disease is mainly found in the UK in May and is thought to persist in any previously infected dried faeces that have been left on the comb.

There are no outward signs of this disease and no viable treatment against this parasite, just good bee husbandry. It does not appear to be that common in the UK; however, it might well be under-recorded.

Insects have been one of the scientists' favoured animals for experimentation, and mainly due to their short life and rapid breeding; we owe a lot to the fruit fly, Drosophila. The bee has also been used by scientists as the Malpighian tubules offer them several options of testing. These organs have been widely used in ecotoxicological studies as they quickly respond to changes in the insect's environment, and these changes, in particular pesticides, can induce the expression of detoxification genes in the cells of the Malpighian tubules. A large number of scientific manuscripts around the world indicates that the Malpighian tubes can be used for the evaluation of biomarkers, detoxification, and innate immunity.

Expansion vessels

The worker honeybee has an amazing capacity for storage in two of its internal organs- the crop or honey stomach and the rectum, both of which are capable of great distension and retention.

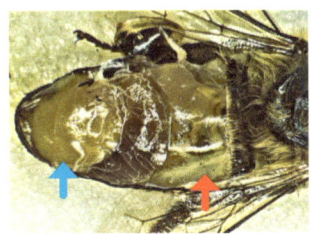

Fig 168. Worker taken from a swarm, showing the crop full with honey, red arrow. The rectum is partly full of liquid faeces, blue arrow. X 10 Magnification.

The crop is a transparent bag-like structure situated at the foregut and is used for gathering water and nectar plus honey when a colony is about to swarm. It is capable of holding up to 30% of the bee's weight with nectar. Dade quotes that the maximum capacity of the crop is 100 mg, although 20 to 40 mg is more normal when foraging. It is estimated that to fill a 454-gram jar of finished honey takes between 12,000 to 24,000 journeys.

The crop is not a true stomach but rather a social stomach as its contents can be exchanged (trophallaxis) with other bees. At its rear end, it has a filter mechanism, the proventriculus, which filters out any pollen grains from the nectar, before it enters the true stomach the ventriculus in the foraging worker bee; pollen is a protein used mainly for growth and is therefore of minimal use to a foraging bee. Young house bees, newly hatch ingest pollen to turn into brood food which contains both protein and carbohydrates to feed to larvae.

Fig 169 (left). The crop from the swarming bee above, dissected out, full of honey. X 40 Magnification.

Fig 170 (right). Normal-sized crop. Red arrow. X 10 Magnification.

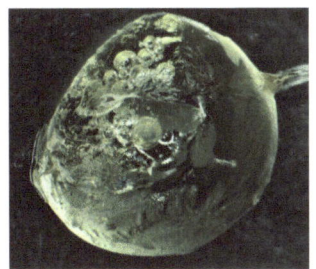
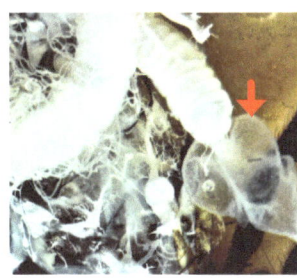

The rectum is part of the hindgut into which the contents of the small intestines empty and from which wastes pass out through the anus. Its role is to hold the waste matter from digestion which is not normally expelled within the hive-bees retain it until they can fly outside in good weather. During the winter months or prolonged spells of inclement weather, they are capable of expanding the rectum into the whole abdomen.

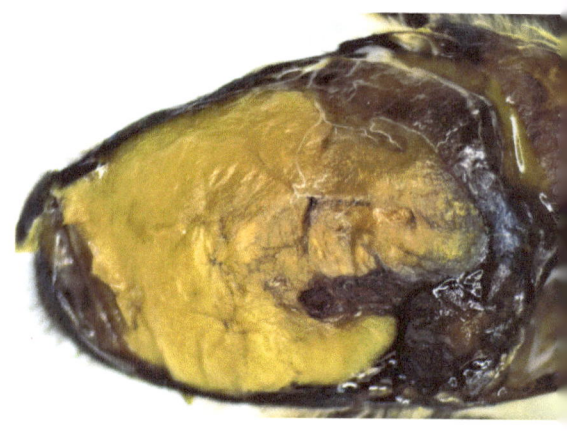

Fig 171. An overwintered worker bee, showing extended rectum. X 40 Magnification.

There are six partly chitinized pads arranged around the rectum, called rectal pads, the role of which is to reabsorb ions and water from the rectum that were previously collected by the Malpighian tubules. They are thin flattened tube-like structures. These work by osmotic pressure.

Bees do not store water in the colony, so must make full use of the water in their bodies.

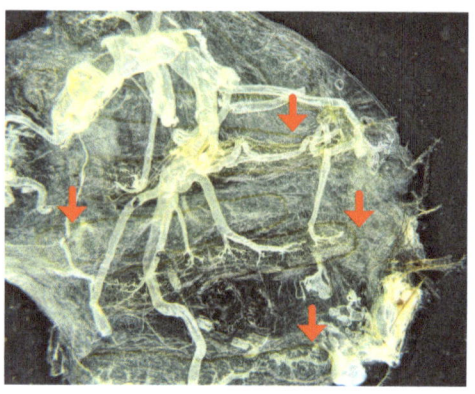
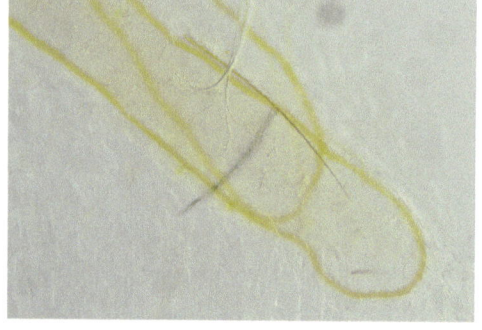

Fig 172 (above). Rectal pads naturally coloured yellow. X 40 Magnification.

Fig 173 (above right). Rectal pads naturally coloured yellow. X 200 Magnification.

Fig 174 (right). Dissection of the worker. Taken from Dade by kind permission of IBRA.

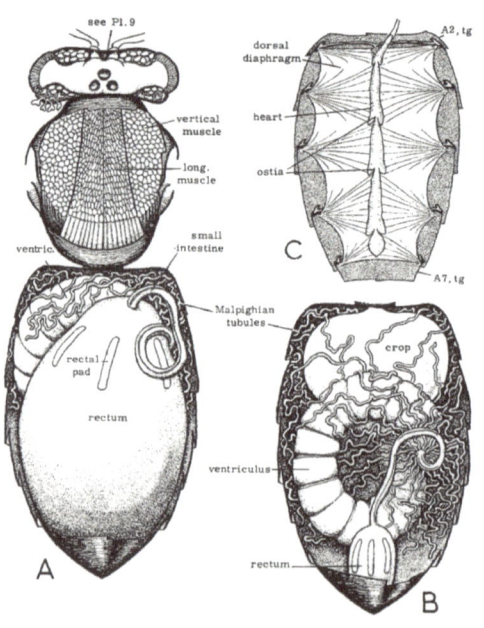

Digestion

Digestion (noun.)

Late 14c., *digestioun*, "conversion of food to a state in which it can be absorbed into the blood from the alimentary canal," from Old French *digestion* (13c.) and directly from Latin *digestionem* (nominative *digestio*) "digestion, arrangement," noun of action from the past participle stem of *digerere* "to separate, divide, arrange," etymologically "to carry apart," from *dis-* "apart" + *gerere* "to carry".

Mandibles are made for many things

The honeybee has two opposing spatula-like mandibles attached to its head; they differ between the queen, worker and drones and form specialised duties in each.

Made of chitin, they are hardened by being sclerotized; this allows them to be extremely strong. Older worker bees often have saw-like notches at the edges, owing to wear and tear. The outside edges have long hairs that overlap; these are innervated and are believed to be mechanoreceptive, sensing biting action. Other smaller hairs that hang over the jaws are also sensitive to movement.

The workers are the specialist's manipulators. Their first job is to mould wax for the comb with their mandibles within the hive. A second role recently discovered is to bite the Varroa wherever it can! Take a look at the damaged ones on your hive floor.

The main job of the young workers between five and fifteen days is to feed the larvae and queen, which they accomplish by producing brood food from the head glands and secreting it down a groove on either side of the internal surface of the mandibles. On venturing outside of the hive the worker uses her mandibles to collect propolis, which is deposited on her back legs; on returning to the hive the house bees will take off the pollen and propolis with their mandibles and store it in the cells or seal up a crack. They are also used to support the tongue when drinking or collecting nectar.

The queen has no major need of her mandibles; they are toothed with a single cusp at the far end, and this is used to dig her way out of her queen cell. The outside of her mandibles has more hairs than the workers.

The humble drone has a smaller set of jaws and like the queen, they are also cusped at the bottom and covered in even more hairs. It is believed that they serve no major role for the drone, other than helping to remove the capping when it emerges from its cell.

Fig 175. The mandibles supporting the tongue whilst feeding-drinking. X 40 Magnification.

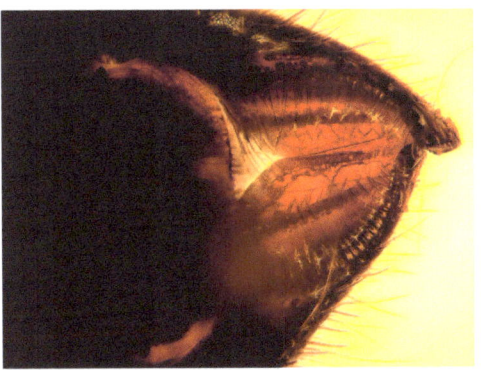

Fig 176. View from beneath the mandibles, showing slight overlap, allowing one jaw to slide inside the other. X 100 Magnification.

Fig 177. Inside view, showing concave effect and grooved channel to allow brood food down from head glands. X 40 Magnification.

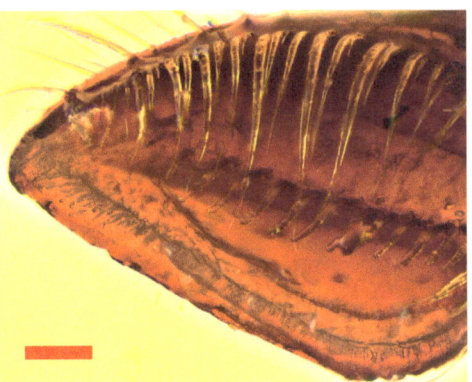

Fig 178. Specialist hairs overhanging the mandible. Scale bar 100 microns. X 100 Magnification.

Fig 179. Drone mandibles, which are much hairier than the worker is. X 40 Magnification.

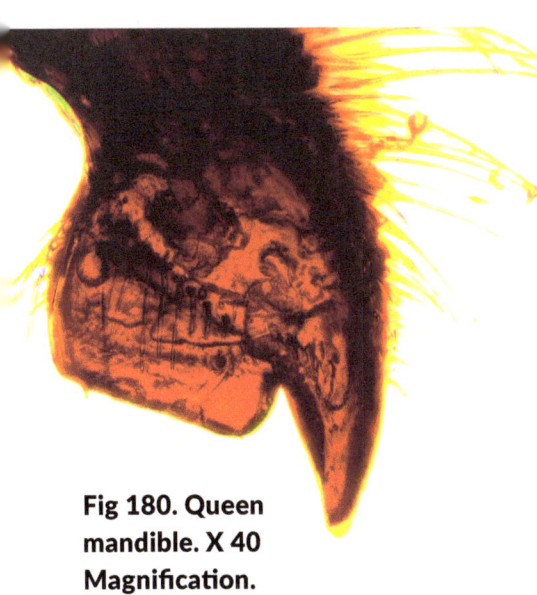

Fig 180. Queen mandible. X 40 Magnification.

Proboscis, a drinking straw by another name

The lower mouthparts of the bee make up the proboscis, 'a feeding tube', composed of the following parts, the stipites, 'a stack or trunk', galeae, 'a helmet', laciniae, 'a flap or fringe' and the vestigial maxillary palps, 'a feeler' on the lower jaw. All of these parts come together to form a tube-like structure used for delivering saliva down the middle tube and for then sucking water or nectar up via a second surrounding tube, the food canal.

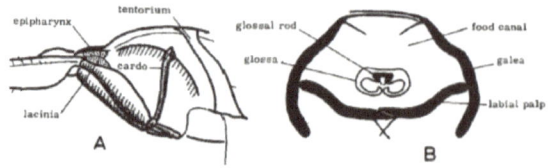

Fig 181. A. The proboscis extended forwards, cardines swung backwards, laciniae pressed against the epipharynx. B. Cross-section of proboscis rolled up, showing the two galeae, overlapping at the top. Taken from Dade with kind permission of IBRA.

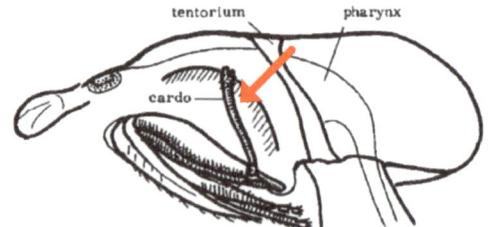

Fig 182. The proboscis folded back, cardines swung forward. The red arrow shows the Cardo, which acts as a lever, moving the proboscis backwards and forward. Taken from Dade with kind permission of IBRA.

Underneath the head, the stipites are joined together by the traverse lorum, 'a thong', and they held between the cardines plural of cardo, 'a hinge'. At the ends they are fixed to pegs in the inner walls of the fossa, 'a trench'. These parts form the mechanism that enables the proboscis to move backwards and forwards.

The inner parts of the tongue all belong to the labium or lower lip. They are the postmentum, meaning behind the chin, which is fixed to the middle

Fig 183. A cleared specimen. Dissected head of a worker bee, showing cardo, white arrow.

of the lorum and the prementum. The labial palps, the long glossa, (the actual tongue) and two paraglossae, meaning alongside the tongue, at the end of the glossa, a small rounded, mop-like flabellum, (meaning a little fan) is situated.

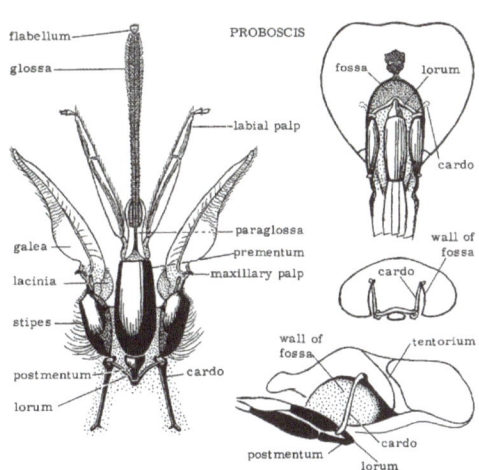

Fig 184. Dissected mouthparts in displayed position.

Fig 185. Proboscis spread out. Taken from Dade with kind permission of IBRA.

The glossa forms part of a hollow tube when flattened and curled at its side; in section, it resembles a C shape. It is stiffened by a rod that can be drawn backwards by muscles inside the head. The glossa is covered with small hairs; these become erect when the walls of the glossa are stretched. The labellum has a fringe of stouter bristles.

The bee produces saliva - a mixture of liquids secreted from the postcerebral and thoracic glands; It is stored in the salivarium a pouch just in front of the proboscis. From here, the saliva runs down the extended tongue between the paraglossae and down the glossa. The glossa acts as a kind of scrubbing brush when used with saliva that has been deposited on a dry sugary substance.

The bee needs to keeps its tongue folded back when not in use. Due to its specialist lifestyle, it has to be able to feed the larvae by using the grooves in its mandibles. She also has to use her mandibles for biting and manipulating wax for comb building. The parts in front of the prementum and stipites are kept behind on the cardines and are tucked away under the fossa.

When in use it is swung forwards on the cardines, which act rather like a lever and at the same time it is unfolded extending forwards employing a combination of internal musculature in the head, and haemolymph pressure. The proboscis is now level with the mouth and once it is clasped by the mandibles, the laciniae are then pressed against the epipharynx, forming an airtight seal.

Fig 186. Drawing of the formed proboscis showing section at the top, the channels. By kind permission of D G Mackean www.biology-resources.com

When the galeae and labial palps are brought together to form a tube around the glossa, the outer curved flaps of the galeae overlap each other to make a seal, creating a food channel to suck up liquid nectar and this uses all of the proboscis parts. By expansion of the ciborium, (Latin for food) in the front of the mouth, it can draw fluid upward through the tube by suction. The bee has another trick it can use to rapidly inflate the glossa. By pumping blood under pressure into the glossa, dust and pollen on the outside can be forcefully ejected, helping to keep it clean.

The drones have tongues about 3mm long as they only have to feed themselves from within the hive stores and not from flowers. On the other hand, the workers have a tongue of about 6mm, which is required to reach the nectaries of the flowers they visit.

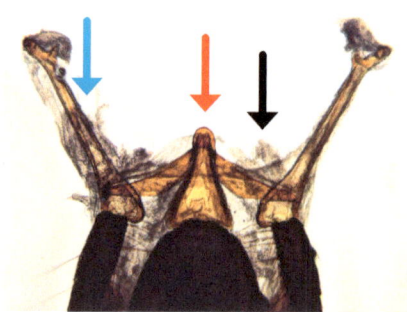

Fig 189. Cardo, blue arrow, lorum, red arrow, postmentum stripes, black arrow.

Fig 187. Worker tongue, about six mm long, in a relaxed position.

Fig 188. Drone tongue, about three mm long, in the forward position.

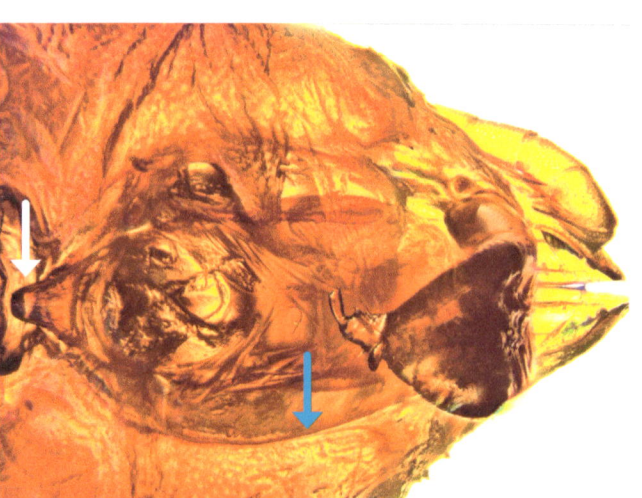

Fig 190 (left). Cleared specimen showing mouthparts from below. Lorum, white arrow. Postmentum stripes, blue arrow.

Fig 191 (above). Worker tongue from underneath.

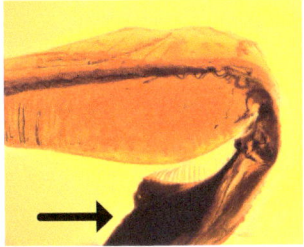
Fig 192. Galea, showing stipes.

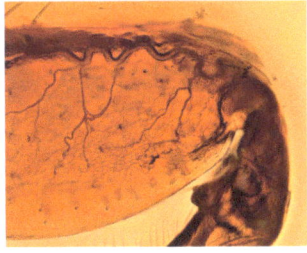
Fig 193. Galea, showing trachea.

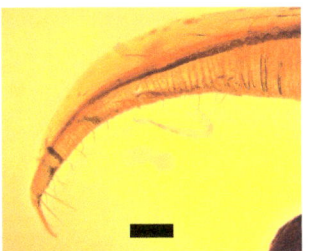
Fig 194. Tip of Galea, showing hairs. Scale bar 100 microns.

Fig 195. Lacinia. Scale bar 100 microns.

Fig 196. Sensory hairs on the end of the labial palp. Scale bar 100 micron.

Fig 197. Laciniae wrapped around glossa. Scale bar 0.5 mm.

Fig 198. Workers mandibles supporting proboscis in the forward position.

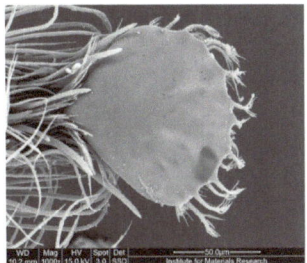
Fig 199. SEM image of flabellum and glossa. © Dr Michel Asperges. Universiteit Hasselt Agoralaan gebouw D te BE 3590 Diepenbeek, Belgium. X 400 Magnification.

Fig 200. SEM image of glossa. © Dr Michel Asperges. Universiteit Hasselt Agoralaan gebouw D te BE 3590 Diepenbeek, Belgium. X 1600 Magnification.

Proventriculus: before the belly

All three honeybee morphs - queen, worker and drone- have a filtering valve, called a proventriculus (meaning before the belly) situated in their honey stomach, also known as the crop. This acts as a storage organ for nectar and water and is capable of great distension. The filtering is done by a mouthpiece that sticks up into the crop from the end of the ventriculus this prevents the contents of the crop from running freely into the true stomach, the ventriculus.

Both crop and proventriculus have an outer layer of transverse musculature and an inner layer of longitudinal musculature. The longitudinal musculature of the proventriculus is powerful and by contraction causes the lumen (cavity) of the organ to enlarge. The 'lips' are extensions of the folds beyond the encircling band of transverse musculature. The combs of filiform (thread-like) hair on the edge of the folds, are capable of being folded in upon themselves and are capable of opening away like a pair of lips.

These hairs give unrestricted entry to pollen grains rushing into the expanding lumen of the proventriculus when the bee requires food; when acting as a particle filter, they are capable of filtering particles as small as one micron. The contraction of the circular muscles will cause the expulsion of the fluid contents of the proventriculus back into the crop, the thicker sphincter layer preventing entry into the ventriculus. Pollen grains will be sieved out by the comb and forced into the pouches as the folds collapse upon each other. Repeated intake and expulsion of contents in this manner will gradually cause a mass of pollen grains to accumulate in each pouch until once a large mass of pollen grains is collected. The contraction of the circular muscles forces this large bulk against the hairs of the combs.

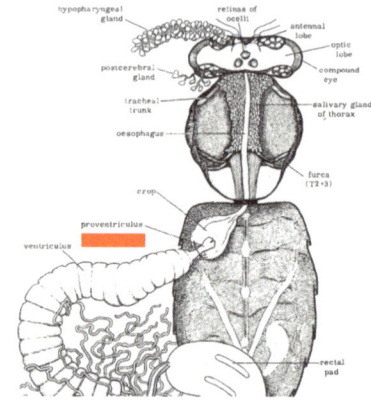

Fig 201. Horizontal section of the worker. The alimentary canal displayed. Taken from Dade with kind permission of IBRA.

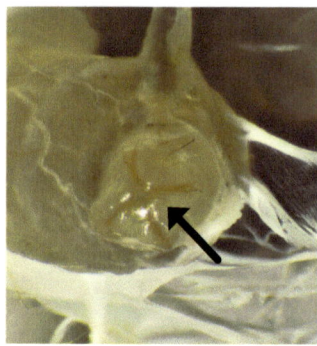

Fig 202. The proventriculus lips showed inside the crop. (The crop has been removed) X 40 Magnification.

Within the ventriculus, the swallowed pollen is kept within a membrane, named the peritrophic membrane. These thin membranes are secreted by the cells lining the ventriculus (epithelial cells). Nearly all insects have similar membranes to these, and considerable study has gone into establishing the role they play in digestion. In the honeybee - these membranes are produced by some of the epithelial cells and successive sheets of membrane peel away from the wall, coating each bolus (a ball) of pollen as it arrives. The membranes were previously thought to protect the lining of the ventriculus from sharp points on the pollen. However, it is more likely that they are important in concentrating a range of digestive chemicals (enzymes) to where they are most needed.

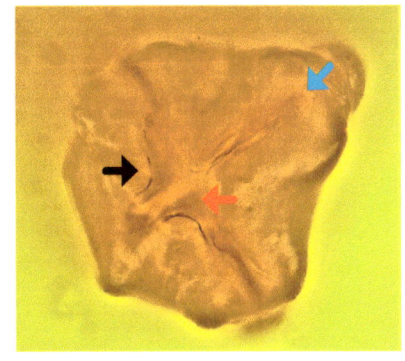

Fig 203. Proventriculus, showing lips, black arrow. Pouch, blue arrow and lumen, red arrow. X 100 Magnification.

The bee scientists, Whitcomb & Wilson, showed that the shells of the pollen grains are not broken at any stage, yet their contents are completely digested in the ventriculus. Within five to twenty minutes, depending on the concentration and amount of pollen suspension, the pollen wrap is passed towards the posterior end of the ventriculus, then on to the intestine, it has been estimated that between 0.15% and 0.43% of the pollen is left in honey.

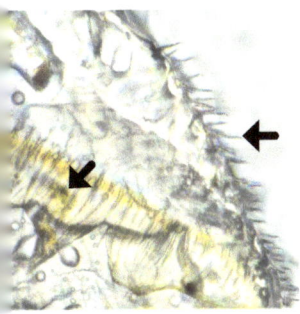

Fig 204. The proventriculus showing the inside view of the lip. X 400 Magnification. Showing long filiform hairs about 70 microns long and bristles.

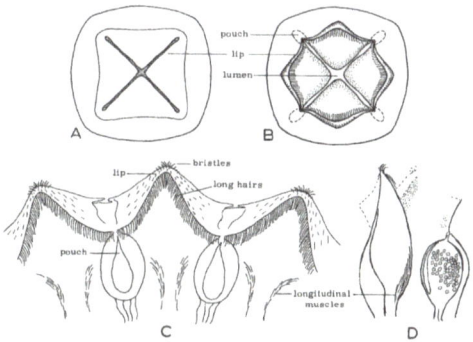

Fig 205. The proventriculus. A. anterior aspect, lips closed. B. lips open showing short spines and long hairs of lips, and the lumen partly closed by the muscles below the lip. C. Part of the proventriculus laid out after slitting up one side, three of the four lips are sown, with pollen pouches between them. D. Sketch of the longitudinal section, on left through a lip, on the right through the pouch. Taken from Dade with kind permission of IBRA.

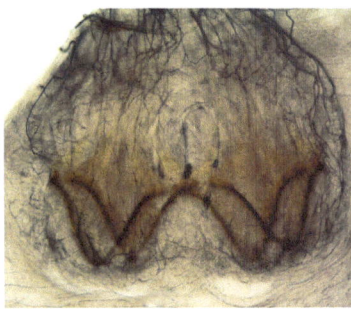

Fig 206. Four lips displayed. © Dr Michel Asperges. Universiteit Hasselt Agoralaan gebouw D te BE 3590 Diepenbeek, Belgium.

Revenge?

I noticed over 60 mustard coloured oval spots about 1 mm in size on my car early in the morning that looked suspiciously like those seen outside the hive when bees have a problem, such as dysentery.

If a bee is unable to void the contents of its distended rectum for an extended time such as in the winter, or due to bad weather, indigestible foods like starches may ferment due to bacteria, yeasts and microfungi causing dysentery.

Were they seeking revenge I wondered? Butterflies were not in evidence en mass, so they were excused from being the culprit.

The day before I had taken a quick look inside the hives for the first time that year and collected 30 bees, mainly from outside the hive (I leave a seed tray underneath the entrance to keep an eye on what is happening without looking inside.) A check for Nosema and Acarine mite produced good news;-none was found in either hive.

I scraped some of the material off the car, put it into some alcohol, and then made a slide with some added stain to help show up the contents. The evidence was conclusive. Guilty as charged. Bee poo!

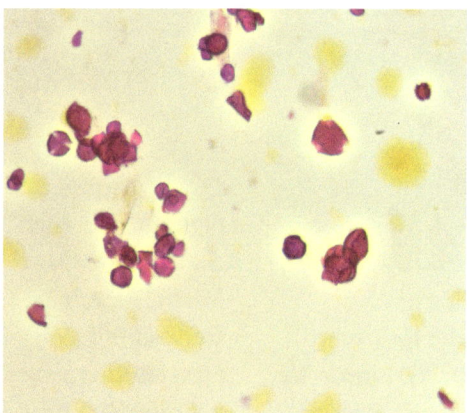 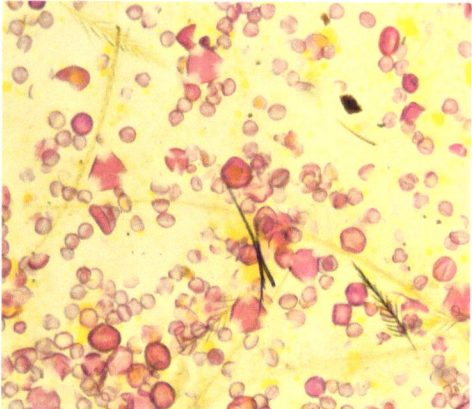

Fig 207. Mustard spot evidence. Pollen stained red also showing the yellow wax from their outer coating. X 200 Magnification.

Fig 208. Comparative evidence. The Nosema sample (none evident). Taken from the ground up abdomens, showing pollen and hairs. X 200 Magnification.

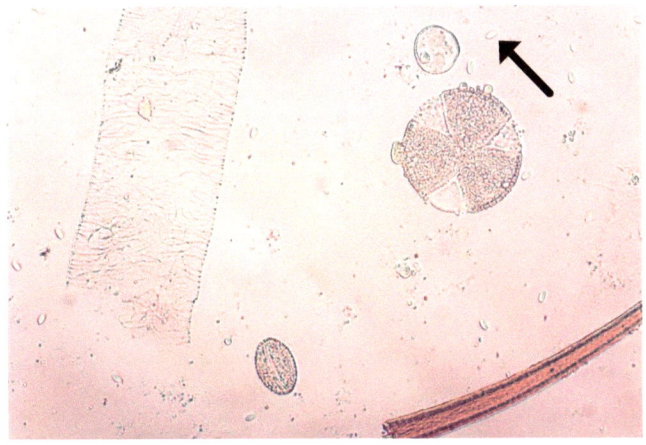

Fig 209. A sample with Nosema, small rice-shaped particles, black arrow and mixed pollen grains. X 400 Magnification.

Whilst inspecting the hive I inadvertently pulled apart some grubs when splitting the super, so I put them under the microscope to examine their gut contents (a mustard coloured liquid) to see what they had been fed. The larvae were about 5-7 days old, so fully formed.

The slide revealed pollen, which larvae need a lot of as it is their only source of protein, fat, vitamins and minerals that fuel their massive growth over those first seven days.

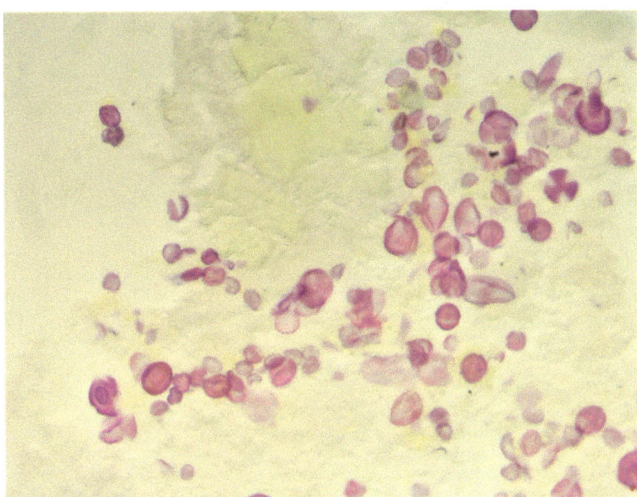

Fig 210. Larvae cut contents. X 200 Magnification.

Faecal matter includes pollen exines, the outer hard coating, the inside cytoplasm being digested, pollen lipid globules, the yellow matter in the pictures, and sloughed epithelial cells from the ventriculus which covers the pollen mass during digestion, this is thought to protect the gut lining from transit damage from the rough outer exine grains.

So there you have it, a potted history of bee poo!

Reproduction

Reproduction (noun.)

1650, "Act of forming again," noun of action from reproducing. Of the generation of living things, from 1782; of sounds, from 1908. Meaning "a copy" is from 1807.

Honeybee drones - possessors of some very specialised equipment

Honeybee drones have similar sex organs to other bees; however, these differ in shape at eversion. This species-specific design stops them from cross mating with other bee species. It has also been noted that when drones from different races within the same species congregate at mating sites, they seem to prefer to mate at different heights, although no one seems to know why.

Fig 211. Drone.

The endophallus, meaning a 'penis held within', consists of a large complicated tube inside its abdominal cavity. In order to ejaculate semen, the drone must first evert it outside of its body. This is achieved via the opening at the tip of the abdomen, the phallotreme, which is also the exit for the anus.

At the proximal end is the thin tube known as the cervix, behind which is a ball-shaped chamber, the bulb, which contain the semen and mucus, this must pass through the cervix at stage two of eversion. On either side are two yellow horns, the mating sign. At stage two of eversion the hairy pad, the vestibulum, is exposed, this acts as an aid for the following drone he has to remove the mating sign left by the former.

There are also specialised small claspers on either side of the opening. These do not seem to offer any assistance during mating in honey-

Fig 212. The drone's external opening, the phallotreme, the arrow shows the claspers.

bees so their function is unclear. In other insects, they aid in mating and clasp onto the female. Being quite small in the drone, they might act as a support for the endophallus as it leaves the phallotreme.

There are no muscles or nerves in the endophallus. It is everted by muscular pressure, which forces the haemolymph (bee's blood) and air, under pressure, to evert the endophallus in two stages. This pressure also forces semen to be ejaculated at pressure into the queen's ovary ducts. The pressure, normally applied by internal musculature during the mating process, can also be generated externally by the beekeeper, for instance between finger and thumb, with a similar effect. Thus, a captured, sexually mature drone will evert its endophallus in two stages presenting sperm on the end, which can then be collected for use in instrumental insemination. An audible pop can often be heard due to the intense pressure involved in eversion.

The black arrow showing the tube, the cervix (meaning neck), which is inserted inside the queen before the second stage can take place.

It is through a tiny space in the cervix that the bulb seen above must pass through. Note the yellow colouring on the two horns. The function of these is described below.

Fig 213. This shows the first stage of eversion, the white arrow showing the dorsal plate of the bulb, a structure within the bulb that keeps it rigid.

At the commencement of the second stage, the end of the tube is inserted into the queen's bursa copulatrix (the mating pouch) which doubles as the sting chamber, and then the second stage full eversion begins.

Note that in mating the bulb at the end would already have been inserted inside the queen; however, it is only the tip, the cervix, which is inserted first, before it becomes fully inflated inside the queen's mating chamber. It is at this stage the mating sign from a previous

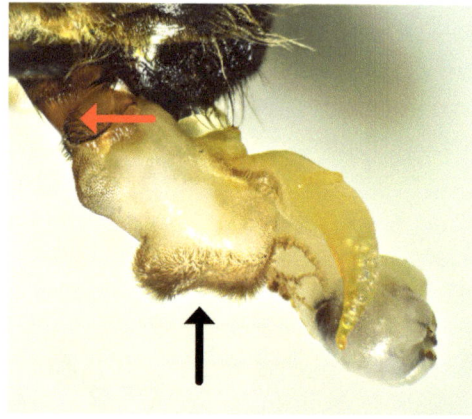

Fig 214. The second stage of eversion of the endophallus. Red arrow showing claspers.

mating is forcibly removed by the final eversion of the mating drone. (Note the arrow position, showing the vestibulum, see below for an explanation).

Mature drones can only mate once their sperm are fully developed - about 12 days post-emergence from the cell. One way to determine if the drone is sexually mature is to examine the two horns, the cornua that appear after the first eversion. They form part of the mating sign that is left behind after mating and act as an attractant to other drones who see them fluorescing in ultra-violet light. Immature drones do not have the yellow coating; hence, these specimens would be discarded if manually collecting semen.

To mate with the queen, the remaining mating sign from the previous drone must first be removed, as explained above. This occurs when the tip of the cervix is inserted and the second eversion takes place, forcing the signal out. The current thinking is that the hairy pad, the vestibulum, shown in figure 214 acts as a brush-like collecting mechanism to help remove the sign.

Mating takes only a few seconds per drone. They flip over backwards at the final stage of mating, leaving behind their mating signal, falling to the ground where they die, having successfully passed on their genes.

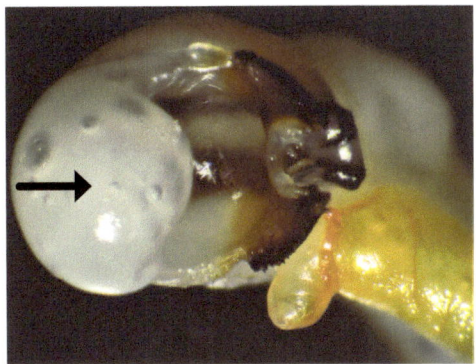

Fig 215. The tip of the bulb, showing semen and mucus.

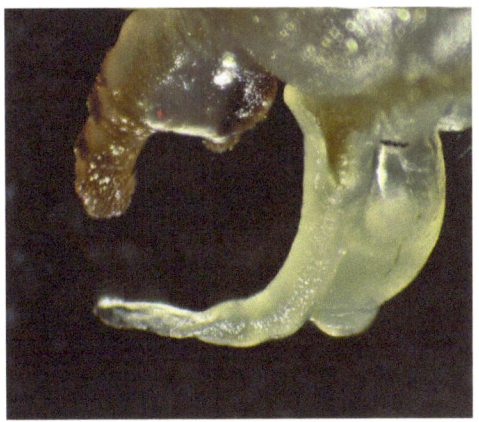

Fig 216. Part eversion of an immature drone, showing the horns lacking in an orange membrane.

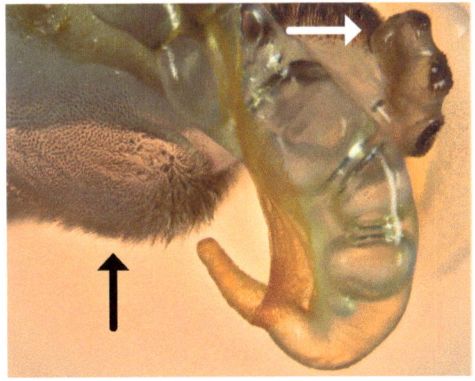

Fig 217. Close up of the brush at the first stage of eversion, black arrow. Tip of the cervix, showing proximity to the brush where it is ready to be inserted inside the queen, white arrow.

The life of the sperm

The sperm, Latin for seed, is the male reproductive gamete, meaning husband in Greek, or wife if it is an ovum. Animals produce motile sperm with a tail, known as a flagellum, meaning a little whip, which is used for movement; the correct anatomical name for a fully developed sperm is spermatozoa.

The life of a sperm's development starts early on in the larval stage. The testes, which make the sperm, take up a large area in the abdominal end of the larva. The testes are made up of about 200 tubules, and the sperm develops at the upper ends. They continue to develop through the prepupal stage and pupal stage and reach their maximum size and maturity when the drone hatches out of the cell.

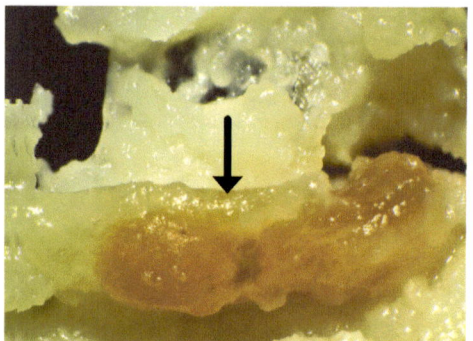

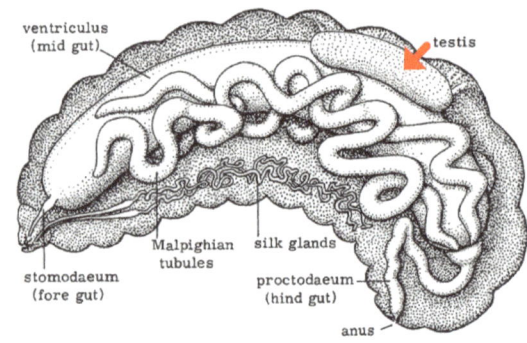

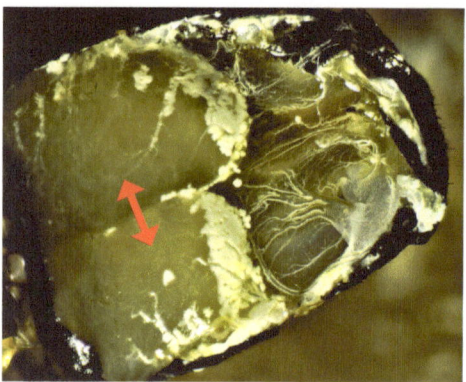

Fig 218 (above left). Testes of a six-day-old drone larva.
Fig 219 (above). Showing testes. Taken from Dade with kind permission of IBRA.
Fig 220 (left). Mature testes of a recently hatched drone.

Between three and eight days after hatching, the mature sperm migrate to the storage vessel, the seminal vesicle, after which the testes shrink down in size and no longer produce any sperm. Their role has been completed. This normally takes between nine to twelve days. Only then is the drone ready to mate. There is a variation in these times dependant on genetics and the colony state of health regarding food availability.

Seminal Fluid, a Potential Problem.

Seminal fluid is a complex mixture of proteins and metabolites with multiple functions. It keeps sperm alive and motile, protects against pathogens, and regulates sperm capacitation - the final maturation step that enables sperm to fertilise eggs.

When females mate with multiple males, some seminal fluid components can become agents of sexual selection and harm rival ejaculates while others manipulate female physiology to enhance a specific male's reproductive success.

Seminal fluid has been shown to cause rapid deteriorating vision in queens, thus reducing their likelihood of leaving the hive the next day to mate again. The queens try to counteract these effects by leaving for mating flights sooner, thereby increasing offspring genetic diversity and the success of their colonies.

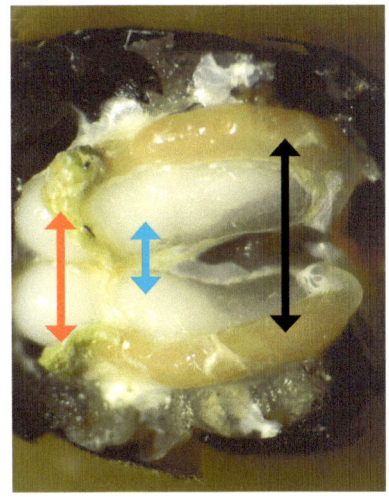

Fig 221. Testes that have shrunk down in size, red arrow. Seminal vesicles where the sperm are now stored, black arrow. Mucous glands, blue arrow.

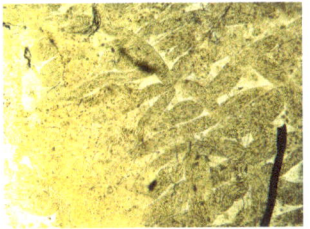

Fig 222. Sample of tubules from testes of recently hatched drone. X 100 Magnification.

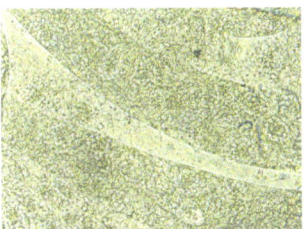

Fig 223. Section through testes tubules. X 400 Magnification.

Fig 224. Ejaculated sperm on the end of the endophallus. This can be collected for artificial insemination.

The queen has a special storage organ for sperm, the spermatheca, which is circular and surrounded by a mass of tracheal netting. The mixed sperm from multiple mating can spend up to five years inside the spermatheca, being kept alive and in good condition. The young queen will allow a few sperm access to fertilise the egg, although only one will succeed. As she gets older, more sperm are released as the egg passes. Eventually, the queen will fail if left to her own devices - often four to five years after mating provided that supersedence has not occurred.

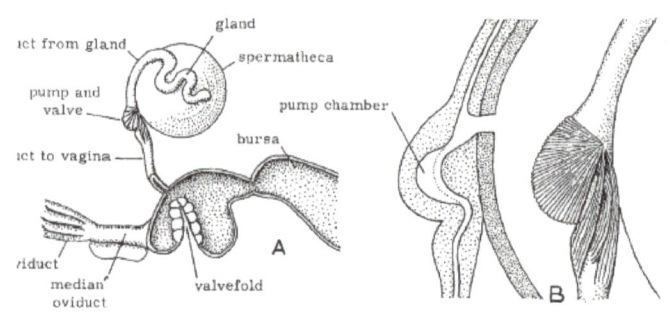

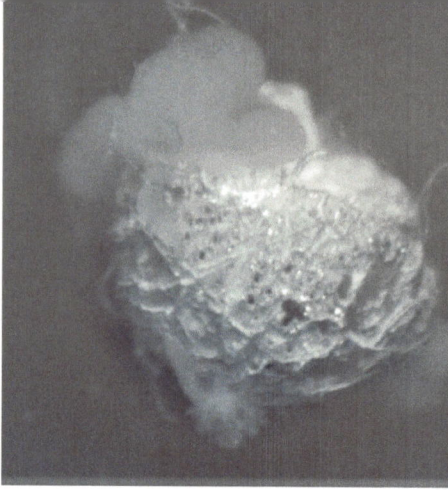

Fig 225. The spermatheca and vagina, with adjoining organs of a queen. B The spermathecal valve and pump, in section, an external view, showing muscles. Taken from Dade with kind permission of IBRA.

Fig 226. The spermatheca showing tracheal netting X 40 Magnification.

Due to polygamous mating, the honeybee queen will mate with up to twenty drones in succession. Recent research has challenged this figure with evidence of up to seventy-seven drone matings. The sperm is forced into her oviducts during ejaculation, where it mixes. After the queen returns to the hive, about 3-5 % of the two to seven million sperm received during mating is stored.

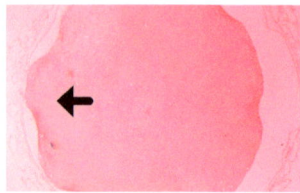

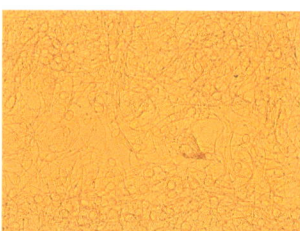

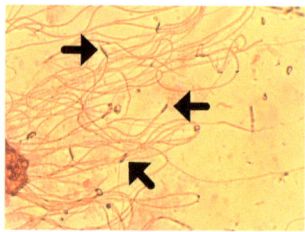

Fig 227. Section through spermatheca, showing a mass of packed sperm. X 400 Magnification. By kind permission of DARG.

Fig 228. Mass of coiled sperm, taken from a two-year-old queen's spermatheca. X 400 Magnification.

Fig 229. Sperm showing heads. X 600 Magnification.

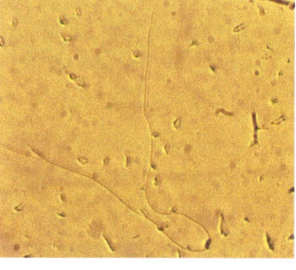

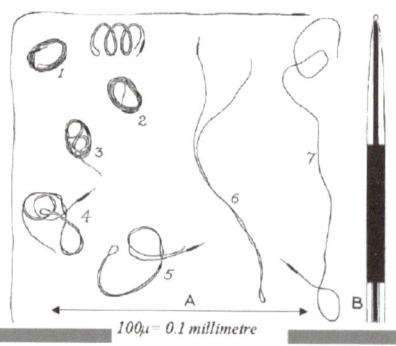

Fig 230. Single sperm cell. X 600 Magnification.

Fig 231. A. 1-2 coiled. 3-7 stages of uncoiling. B. Showing structure of head and part of the tail. Taken from Dade with kind permission of IBRA.

Anatomy of a Sperm

Sperm take on a slender thread-like shape with a small head at the end terminating with a short spine. The tail is about 250 microns long and 0.25 mm by 0.5 microns in diameter; with a head about 10 microns long.

The head contains the nucleus containing the 16 chromosomes, with densely coiled chromatin fibres; these play important roles in reinforcing the DNA during cell division. The outer surface of the head is surrounded by a thin layer called the acrosome that contains enzymes used for penetrating the female egg. In the honeybee - it has a spike at the front. See the Dade image above. The tail, the flagellum contains 9 fibrils, minute fibres and twelve mitochondrial strands. It is the longest part of the sperm and is capable of spiral-like movement that helps propel sperm towards the egg. The midpiece has a central core with many mitochondria spiralled around it. These act like a storage battery for energy for propelling the tail. The primary sperm cells, the spermatogonia, divide many times increasing their number up to ten million. They grow into large spermatocytes and later are elongated as spermatids and finally, they are transformed into spermatozoa. All sperm are developed during larval growth.

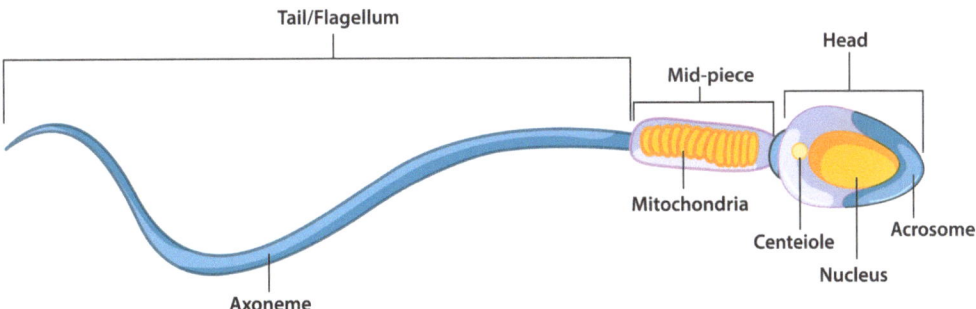

Fig 232. Diagrammatic sectional view of a typical sperm. NB. This is not a honeybee sperm.

In the drone's seminal vesicles and the queen's spermatheca, the sperm lay side by side. When ejaculated they rotate their tails up into three coils but once in the queen's ovary ducts, they straighten out. If the sperm has a shortened tail it will swim in circles. One thought for the extremely long tail is due to their corkscrew-like movement rather than wiggling from side to side. The sideways movement favours swimming in liquid but due to the minute size of the queen's ducts, the sperm tails have to push against the sides to propel themselves forward.

Nature's production line

Fig 233. The ovaries from a four-year-old queen, white lines are trachea. X 20 Magnification.

The queen bee has often been credited with laying up to a maximum of 2000 eggs a day in the latter days of spring. This equates to laying a third of her body weight in eggs, so just how does she do it?

Her main reproductive organs are the ovaries that sit on top of her digestion system. These accommodate up to two-thirds of her abdominal cavity and terminate at a Y shaped joint of the two laterals, with a common oviduct. It is here that the tube from the spermatheca meets the oviduct, it is the storage vessel where the mixed sperm from the multiple mating is held for the duration of the queen's laying life. At the junction of the vagina and spermatheca, there is a protrudance known as a valve fold where sperm is released in front of the egg. As the egg, passes over the valve fold it is either pushed downwards out of the way if the queen decides to release some sperm to fertilize the egg which thus becomes a worker or leave it unfertilized to become a drone. This choice is achieved by the set of muscles at the base of the spermatheca that controls the Breslau's sperm pump.

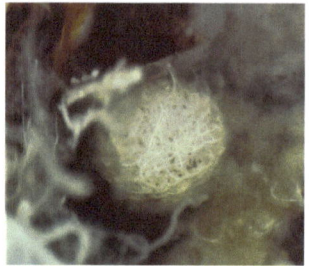

Fig 234. Spermatheca measuring 1.3mm in situ, showing white tracheal netting, this supplies oxygen direct to the sperm. X 20 Magnification.

Honeybee queens possess this specialised structure, named after its discoverer, Breslau in 1905. Located between the spermatheca and the spermathecal duct. It is hypothesised to be important for the transport of sperm to the spermatheca. A recent study suggests another possible function that of continuously sampling a constant volume of sperm and spermathecal fluid from the spermatheca and transferring it to the eggs, thereby helping to restrict the excessive volume of sperm when fertilising an egg.

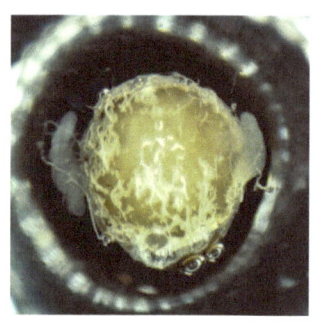

Fig 235. The above picture with some of the netting removed. Pearl-like colour denotes a spermatheca of a fertile mated queen; in a virgin, it would be clear and transparent. X 40 Magnification.

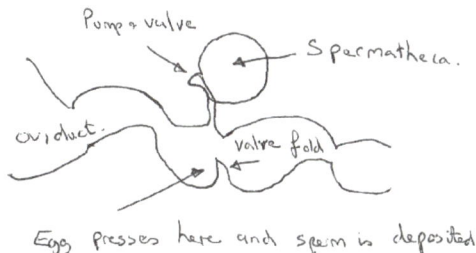

Fig 236. Diagrammatic view of the eggs fertilisation process.

The ovary is made up of as many as 180 long tapering tubes, similar to a string of oval beads with eggs in various stages of development together with nutritive cells. Named ovarioles, they produce up to one million eggs in the lifetime of the queen. The laying worker by comparison has between three and twelve ovarioles.

Fig 237. Compact ovarioles from queen. X 20 Magnification.

Fig 238. Ovarioles, July 2015, Eggs in varying stages of development. X 40 Magnification.

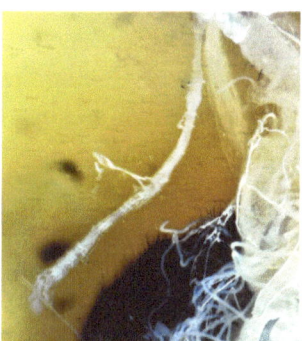

Fig 239. Ovary from a worker. X 40 Magnification.

How eggs are produced

At the top, the thin distal end, of each ovariole there is a multi-nucleate protoplasmic mass so small that the cell boundaries are not defined. From it, germinal cells divide and produce primary oogonia that can be seen under microscopic examination. The oogania divides twice into four cells - One becomes the oocyte, the egg; the other three divides further producing between 32 and 64 nurse cells by mitotic cycles (mitosis is a process where a single cell divides into two identical daughter cells) to become irregular shaped Trophocytes cells.

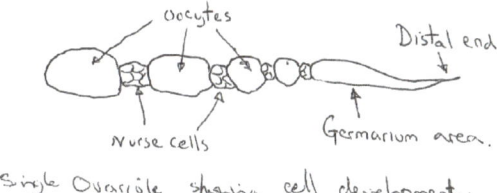

Fig 240. Diagrammatic view of ovariole cell development.

These are the nurse cells that provide all the nutrients needed for development. The egg is also surrounded by a sheaf of follicle cells that are connected at the end by a canal formed by germ cells from which they receive nutrients from the Trophocytes cells and which accompany the egg on its progress. Once these have been consumed, the egg is fully developed. It is near the end of its development that yolk, which contains lipids, protein and carbohydrates, is added to the egg as a source of food. Finally, an outer sheath produced by the follicle cells is added as a form of protection.

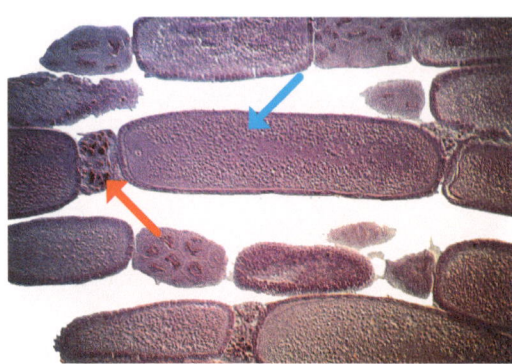

Fig 241. Section through the ovariole showing the finalized egg complete with outer coating, the chorion. Red arrow: degenerating nurse cells; blue arrow: chorionated egg. By kind permission of the Dr Heming University of Alberta, biological sciences.

The final egg is elongate, is slightly tapered backwards and curved in the longitudinal section. At the final developed state, it is about 1.5 mm long by 0.33 mm in diameter and weighs 0.13mg. At the anterior end is a small orifice called the micropyle which is where sperm gain entry to fertilize the ova as it presses against the valve.

All workers have the same mother but some different fathers, making them half-sisters; however, all drones have the same mother and grandfather making them identical.

The math's magic

If the queen lays a maximum of 2000 eggs in a day from the maximum number of ovarioles 360 (2 sets x 180) this equates to one egg being laid every 43 seconds. (24 hours ÷ 2000 eggs = 43 seconds). Could this be why beekeepers can never find their queen? If every 43 seconds she has located an empty cell and enters her abdomen to lay an egg, she is submerged most of the time!

Some average figures can be used to demonstrate the development time for each egg passing down the ovariole from start to finish at - 4.3 hours. (2000 eggs per day ÷ 360 ovarioles = 4.3 per hour)

Flight

Flight. Noun.

Old English *flyht* "a flying, act or power of flying," from Proto-Germanic *flukhtiz (source also of Dutch vlucht "flight of birds."

Late 14th century. from Middle English *fliht*. Sense of "swift motion."

Diaphanous wings

Wing development has been completed by the time a bee emerges from its cell, it only has to unfold and expand them both employing hydrostatic pressure of the haemolymph, and then dry them before they become functional.

Fig 242. Wing of a drone about 23 days old, removed from a cell.

Wing sizes vary between the queen, worker and drone. Drones are much larger and their wings extend beyond the end of the abdomen. The drone is up to two and a half times larger than the worker is. The worker bee's wingtips will lose some of their shape due to wear and tear as she reaches the end of her life, after foraging for three weeks.

Fig 243. Wings of a drone folded back at rest, extending beyond the abdomen.

Fig 244. Worker bee with both wings attached, note how much smaller they are in comparison to the drone.

Fig 245. Worker forewing top, Drone forewing bottom, size comparison.

There are two wings on each side of the thorax, the larger fore wing and the smaller hind wing, the latter of which has lots of small hooks on the top leading edge called hamuli; these hook over the lower edge of the fore wing into a channel when they are first unfolded in preparation for flight, forming a strong bond. The hamuli are spiralled in shape, which helps provide grip and flexibility during flight. The worker bee has between 15 to 27 hooks, the queen from 13 to 23, the drone from 13 to 29. The wing beat can vary from between 200 to 280 beats per second, depending on the caste. Summer midges possibly hold the record for insects; they beat up to one thousand times per second.

If a forager has a flight journey of three miles, it could be accomplished in about fifteen minutes, which equates to a quarter of a million wing beats.

Two sets of indirect flight muscles within the thorax supply the energy to flap the wings up and down only; other smaller direct flight muscles control the feathering and orientation of the wings. The wings are not just flat paddles; they are capable of bending in certain areas to become aerodynamic in shape. The picture below shows the two trailing edges outlined in red; it is beneath these veins that they fold slightly in response to flying conditions. To fly, a wing must generate lift. This occurs when air is forced over the top surface at a faster rate than beneath it. However, another force interferes with lift and drag, namely the resistance of the air over the surface of the wing. It can vary due to the speed or the attack angle, the latter an aerodynamic term for the angle between the oncoming air and the wing.

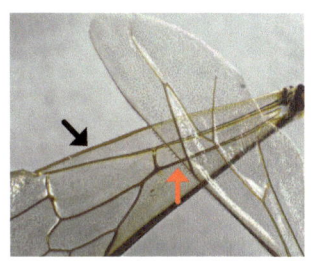

Fig 246. Fore wing at the bottom, showing strengthened leading-edge, black arrow. The hind wing lying across the top with hamuli on display, red arrow. X 10 Magnification.

The wings are covered with small spines on both sides, thought to act as stress sensors and to help break up the friction caused by the drag. (For an in-depth description of honeybee flight see chapter seven of 'Form and Function in the Honeybee' by L Goodman.) The air viscosity affects small insects when they fly; it is akin to flying through syrup or like us walking through water in a swimming pool. The wing vein pattern of longitudinal and transverse veins, called venation, can be used to check on the different species and subspecies of honeybees by comparing certain ratio measurements between two named veins and the discoidal angle shifts between them, together with tongue and hair length. Morphometry is the precise study and comparison of anatomical characters by measurement, while morphology is the descriptive study of form and structure; both words are used throughout the world - for measuring bee bits!

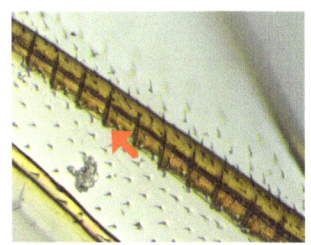

Fig 247. Hamuli engaged and small spines. Scale bar 50 microns. X 100 Magnification.

The wings are made of cuticle membrane supported by a network of veins, some of which are hollow carrying nerves and tracheae to oxygenate the cells, as well as to allow the haemolymph to circulate. The degree of rigidity and flexibility is maintained by the hydraulic pressure within the veins from the flow of the bee's blood, which also helps keep them hydrated and flexible.

In the honeybee, the fore wings are powered and the hind wings, which are attached by the hooks, are mainly following passively. The drones have additional rear wing venation - they have different longitudinal veins, partially branching, and longitudinal veins cross-linking to each other, helping to strengthen the wings in flight. Every vein, depending on its location on the wing, has a certain name; for example, there are four longitudinal veins: costal, subcostal, medial and anal, beginning from the base of the wing.

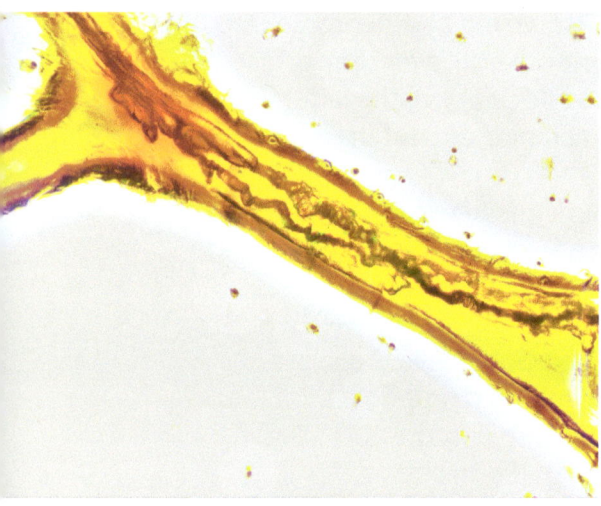

Fig 248. A vein showing the channel for haemolymph and fine tracheae that supply oxygen. X 400 Magnification.

When at rest the wings will be folded back over the abdomen using two bends in the joint area close to the body. This is a complex joint mechanism that helps to provide the sweeping motion of the wing in flight. The wings when at rest also allow the thorax muscles to vibrate without affecting the wing motion; the heat thus generated is used to warm the bee before it flies and, if a house bee, to help in the thermoregulation of the hive.

Fig 249. Wings folded back. X 40 Magnification.

Wings have two non-flight uses - when ventilating the nest, bees grip the surface of the comb and fan their wings to drive airflow through the nest; a similar wing-fanning behaviour is used to disperse volatile pheromones from the Nasonov gland at the hive entrance to help returning bees recognise their colony.

One of the major problems that affect wings is several types of virus - K Wing Virus and Cloudy Wing Virus. However, Deformed Wing Virus, which is carried by the Varroa mite, affects the development of the wings before the bee reaches the adult stage - they emerge with shrivelled wings and are unable to fly.

Morphometry or Morphology?

Morphometry is the precise study of anatomical characters by measurement whereas morphology is the study of form and structure.

Measurement of wing vein characteristics is used to establish race and breeding purity in honeybees, to use the data for selective breeding purposes. This can easily be done by the beekeeper using a variety of methods; I have used a dissecting microscope and a bespoke camera with suitable software. 30 dead bees from 2 hives were collected in early November from outside their hive entrances. One hive was a new nucleus raised in late 2016 from Cornwall and the other a local swarm collected in 2017. Both sets of bees looked much the same in colour. The right forewing was removed from each bee.

Racial types and strains of honeybees have distinctive body characteristics that can help to distinguish both type of bee and the purity of the breed. Of these two aspects, the one of greatest importance is the purity of strain or, more precisely, the degree of hybridisation, (the lower the better). These methods are all of secondary importance to 'colony assessment' characteristics and should be used to refine partly selected strains rather than as a direct descriptor of race. There is no point in propagating 'bad' or undesirable behavioural traits regardless of how 'pure' the strain is.

The phenotype of an individual honeybee, a colony, or a population, is the set of *observable* characteristics such as size, colour, honey production, wintering ability, and defensiveness. It is these characteristics that the bee breeder aims to alter.

The genotype of a bee or colony is the set of inherited genetic instructions encoded in its DNA. These can be either dominant or recessive, depending on the mix from each parent.

Both in natural selection and traditional selective breeding, the selection is applied to the expression of the phenotype, rather than the genotype, since it is the phenotype that directly interacts with the environment and provides a set of observable characteristics.

Measurable characteristics

The general appearance is a simple and obvious characteristic that most would agree on, but it is subjective in nature and one needs to be wary of 'seeing what you desire to see'.

Body colour is only important if describing a bee with a high degree of "purity of strain". Colour in itself is not a positive indicator but can be used to rule out certain hybrids. I have experienced a wide variety of colour in my hives, no doubt from the mixed drones the queens have mated with.

Rings and spots may help give additional information on the degree of hybridisation.

Drone body colour is used as some strains exhibit differences between male and female colouration; this character has more consistency as hybridisation decreases.

Two wing measurements are used to indicate the purity of strain against known data for different species.

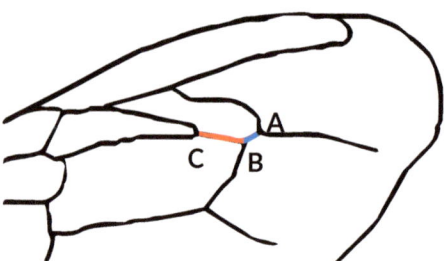

Fig 250. Points A, B & C should be judged to the centre of the vein junctions concerned. The distance "BC" is divided by the distance "AB

Fig 251. Discoidal Shift. This is used to identify Apis mellifera mellifera, which has a negative value, as most other types are zero or positive.

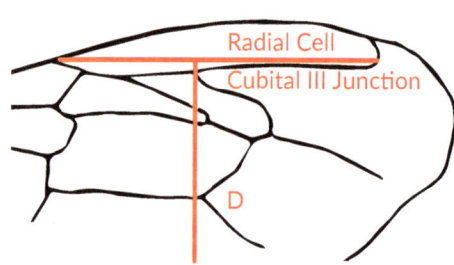

Fig 252.

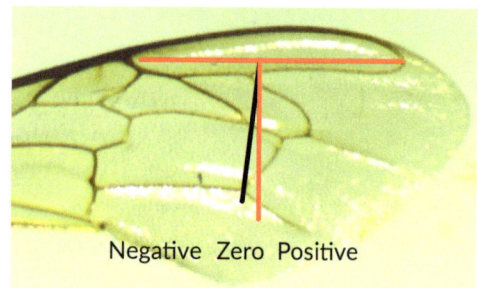

Fig 253.

1. Cubital Index. By measuring the ratio of two of the wing vein segments, we obtain measurements that are consistent for given races of bees; it does not matter what measurement method you use as the result is a ratio. I used the pixel count.

The discoidal shift is an angular measurement in degrees, within the cells of the wing venation. A parallel line is drawn within the radial cell. Then a right-angled line is drawn through the Cubital III junction; from this, a third line can be drawn from the starting point through the Cubital vein and the angle noted. If it is to the left as shown above, then it is said to be negative, central zero, and to the right, positive.

Set out below is an abridged version of the characteristics of each strain, taken from D Cushman's site.

Heavily hybridised bees will show multiple results and not offer any positive results, except to say that is of mixed race.

Character	Apis mellifera mellifera	Apis mellifera ligustica	Apis mellifera carnica
General Appearance	Large, broad, short limbs	Medium size, slim, long limbs	Medium size, slim, long limbs
Worker body colour	Black		Black
Rings		1, 2 or 3 - yellow. Scutellum may be yellow	Maybe one leather coloured ring
Spots	None or small (2nd tergite)		May have small spots
Drone body colour	Dark	Amber/yellow	Dark
Rings or Spots		Yellow rings	Small spots
Cubital Index (worker) average	1.7	2.3	2.7
Cubital Index (worker) min	1.3	2.0	2.4
Cubital Index (worker) max	2.1	2.7	3.0
Discoidal Shift, worker	Negative	Positive	Positive
Worker hair colour	Few dark hairs	Yellowish	Grey
Drone hair colour	Brown/black	Yellowish	Grey or grey/brown

From the data collected, we can see the following results.

Nucleus hive. Discoidal shift angle ranged from zero to 6 degrees. Average value = 2.9 degrees. Ten-registered zero, the rest were positive.

Cubital index ranged from 1.46 to 3.69 Average value = 2.39

Looking at the table, we can assume that we have mostly Apis mellifera ligustica, with a few mixes.

Swarm hive. Discoidal shift angle ranged from two to 9 degrees. Average value = 5.06 degrees. They were mainly positive.

Cubital index ranged from 0.9 to 1.93 Average value = 1.55

Looking at the table, we can assume that we have a lesser hybrid, more Apis mellifera mix.

As my hives are placed close together, I must assume that there must be some 'guests in each colony'; it has been estimated that up to 40% of bees in a hive have drifted there.

Does the swarm consist of wild bees or are they escapees from another hive? A second way to check is to measure the over hairs and Tomentum width (fine matter hairs) in addition to wing measurements as described above.

The abdomens of 30 bees from each hive were separated from the rest of the body and examined under the dissecting microscope. One could also use x 5 optical eyeglass. (At the end of the exercise, save the 30 abdomens and grind them up to check for Nosema)

The tomenta are considered as narrow (less than 50% of tergite), medium at about 50% and broad if more than 50%. The 4th tergite is the segment used to judge the Tomentum width.

The over hair length is judged on the fifth tergite and is compared to a 0.40 mm wide wire.

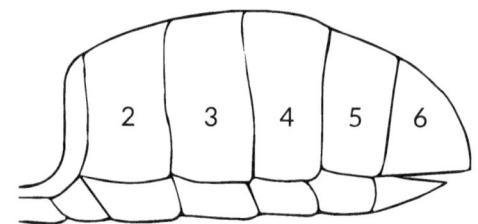

Fig 254. The abdominal tergites are numbered from 1 to 6.

4th Tergite, 2 different bees.
Fig 255 (left) narrow band;   Fig 256 (right) a broad band.

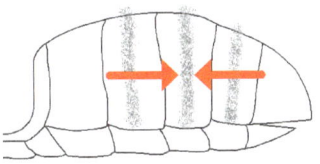 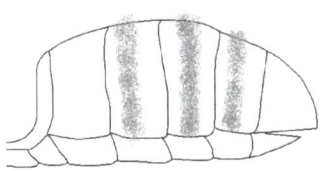 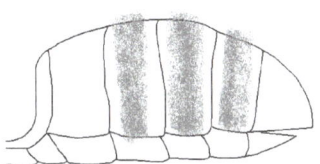

Fig 257. These would be considered narrow tomenta.

Fig 258. These would be considered medium tomenta.

Fig 259. These would be considered broad tomenta.

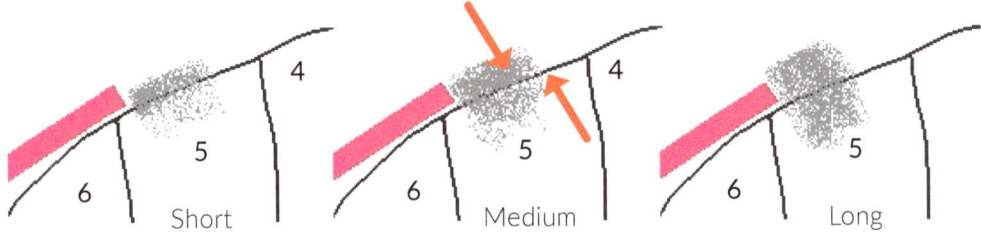

Fig 260. Short over hair, comprises of anything up to 0.35 mm.
Medium over hair fall in the range of 0.35 mm - 0.40 mm.
long over hair, anything longer than 0.40 mm.

Fig 261. This bee has a 0.3mm needle placed nearby for direct comparison, resulting in long over hairs.

Apis mellifera has narrow tomenta and long over hair. This compares with Italian strains that have broad bands of very short hair.

Dave Cushman's data chart

Character	Apis mellifera mellifera	Apis mellifera ligustica	Apis mellifera carnica
5th Tergite Over hairs (mm)	0.4-0.6	0.2-0.3	0.25-0.35
Tomentum Width (4th tergite)	narrow, less than 1/2 of tergite	broad, more than 1/2 of tergite	broad, much hair

A comparison of the two colonies using Dave Cushman's chart provides the following results:

- **Nucleus colony tomenta over hair length average** were 0.3 and broad coverage of hair.

- **Swarm colony tomenta over hair length average** were 0.4 and a narrow covering of hair.

The conclusion using both sets of data is as follows:

- The **nucleus hive** is a mix but shows a stronger tendency towards Apis mellifera ligustica, with a few mixes.

- The **swarm hive** shows strong evidence for being more of an Apis mellifera mellifera mix.

This article has been assembled from several sources, primarily Dave Cushman's excellent web site - www.dave-cushman.net

The Norfolk beekeeper also offers a vast range of short YouTube videos covering a vast array of subjects including microscopy- a must watch! www.norfolk-honey.co.uk/how-to

Muscles

The powerhouse of flight

There are two sets of muscles that power and control the bee's direct flight, an external set of four muscles on the forewing and three on the hindwing. The second set are the indirect muscles that take up most of the inside of the thorax. The forewings are the only ones that have the indirect muscles attached to them; the hind wings are connected to the forewing by hamuli that join them together as they unfold. The action of these indirect muscles causes the backward and forward motion and the primary canting or slanting of the wings.

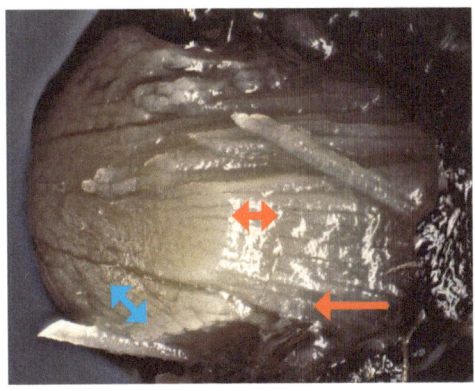
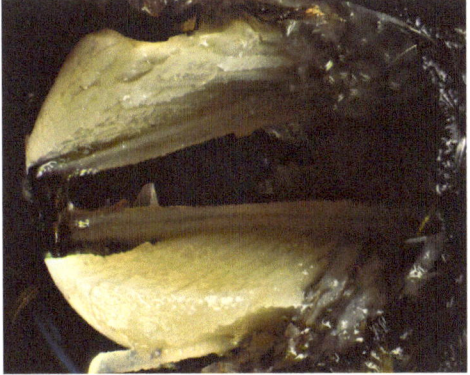

Fig 262 / 263. Left: Preserved specimen showing thorax of the worker bee, showing longitudinal (red arrow) and vertical muscles (blue arrow). Right: The same bee, with longitudinal muscles, removed.

The dorsoventral muscles consist of two bundles situated on either side of the thorax and are attached to the dome of the scutum and the sternite.

The longitudinal muscles are also in two bundles. These run side by side from the tergite and the first and second phragma (Greek for a fence). When the longitude muscles contract and the vertical ones relax, they cause the roof of the thorax to rise, and the wings to depress downwards. When the longitudinal muscles relax and the vertical muscles are contracted, the roof of the thorax is pulled down and the wings are raised. There is a flexible membrane called the scutal fissure that allows the thorax to move.

Fig 264. Drone longitude red flight muscles.

As muscles can only pull, the two sets of muscles must operate alternately, causing the wings to flap.

Their wings beat over a short arc of about 90 degrees and up to 280 beats per second.

Fig 265. A bundle of longitude flight muscles spread out.

The minimum temperature for active foraging is 8 to 10°C. The limiting factor is the temperature of the bee's thorax, which they must be able to keep above 30°C for flight. Bees can elevate their body temperature even though they are ectothermic, (cold-blooded by contracting their wing muscles to produce heat.

The workers often used this heat to warm the comb and larvae.

The thorax muscles in live bees appear red, due to the complex protein, cytochrome in the cell mitochondria, which act as the cell's powerhouse, by converting sugar into energy.

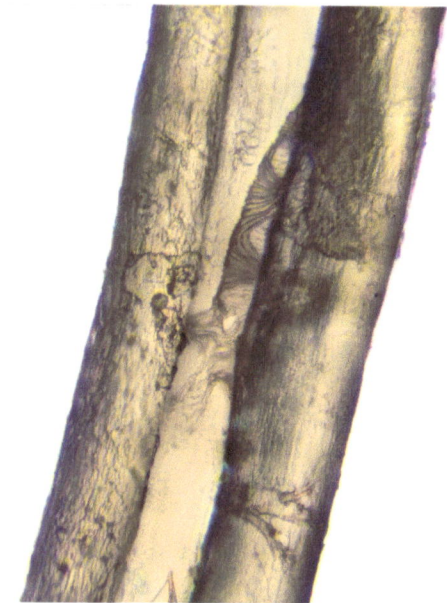

Fig 266. Longitude flight muscle fibres. X 200 Magnification.

The direct flight muscles are individually neurally innervated and contract on receipt of a nerve impulse. The maximum rate of contraction and relaxation of these neurogenic muscles is set by the time taken to receive and act on the impulse message. Once set in motion the indirect muscles will continue to contract and relax until stopped by the nervous system. The muscles are made up of fibrillar, which are large-diameter fibres made from circular myofibrils, which are in turn bundles of protein filaments that contain contrastable elements of cardiomyocytes. Simply put they are the motor that drives the contraction and reaction of the muscles. The dorsoventral (transverse) and longitudinal muscles show myogenic contractility which allows for greater wing beats than ordinary muscles served by nerves, as the maximum beat that the latter can achieve is about 100 beats per second.

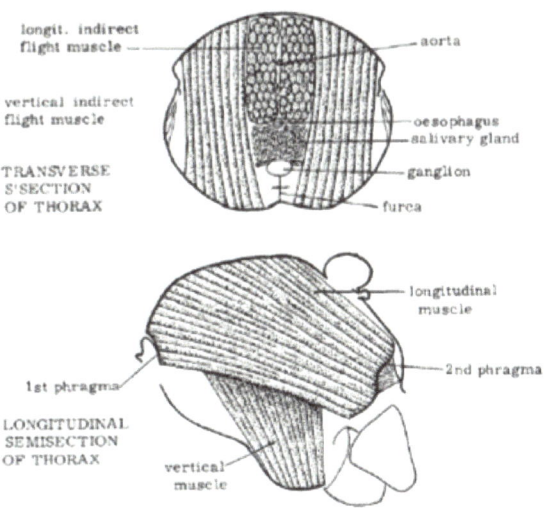

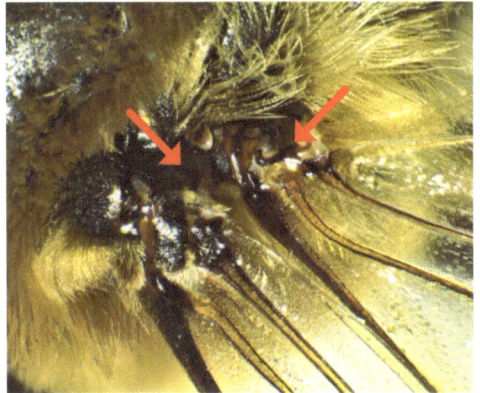

Fig 267. Transverse and longitudinal semi sections of the thorax to show the indirect flight muscles. Taken from Dade with kind permission of IBRA.

Fig 268 (above right). Wings joint in-flight position, the direct flight muscles are attached to the wing sclerites.
Fig 269 (below). Phragma at each end of thorax, muscles removed.
Fig 270 (right). Longitudinal muscles, vertical muscles removed. The red arrow indicated the head end.

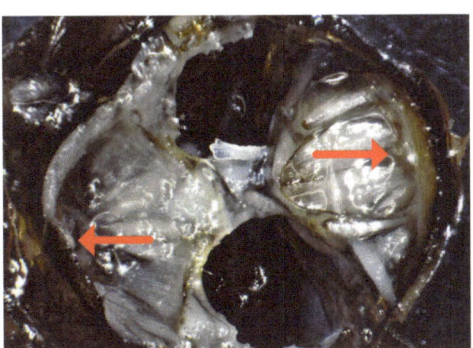

Locomotion

Locomotion (noun.)

1640, "Action or power of motion," from Latin *loco* "from a place" ablative of *locus* "a place;" + *motionem* (nominative *motio*) "motion, a moving" From 1788 as "moving from place to place."

Designed for all events

As the feet of all castes of the honeybee have universal tasks to perform such as locomotion and weight-bearing, it is not surprising that they share a common basic design. However, unique functions particular to the queen, worker or drone also has to be accommodated.

The queen releases pheromones through her feet to advertise her presence in the hive. The drone has a larger, more angular claw than does the worker. The latter has taste receptors on her legs and feet. (The drone seems to lack these although further research might reveal otherwise.)

Sensillae are innervated senses organs that convey different information depending on their type. (This fact was discovered in 2009 by Lorenzo, who used electrophysiological methods to confirm their presence.)

There are high concentrations of taste sensillae on the terminal claw-bearing pretarsus on the forelegs of the worker only.

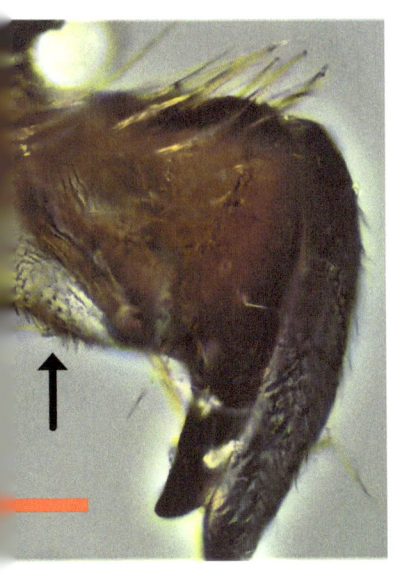

Fig 271. Inside of claw showing the area where taste receptors are located. Scale bar 100 microns. X 200 Magnification.

When honeybees walk over a glass plate, they sometimes deposit from their feet an oily, colourless secretion that has low volatility. When deposited at the hive entrance and on flowers this secretion seems to affect the behaviour of other workers. This pheromone comes from the tarsal gland, also known as the Arnhart gland, situated in the fifth section (tarsomere) of each leg and is found in all morphs although the queen produces much more of this secretion than do the drone or worker.

The method of transmission from the glands to the foot has yet to be conclusively found. The queen produces much more of this secretion than the other morphs.

Recent evidence has found that there is another separate gland around the unguitractor tendon, which is released at the base of the unguitractor plate on the foot. This tendon acts on the ungues (claws) enabling them to grip.

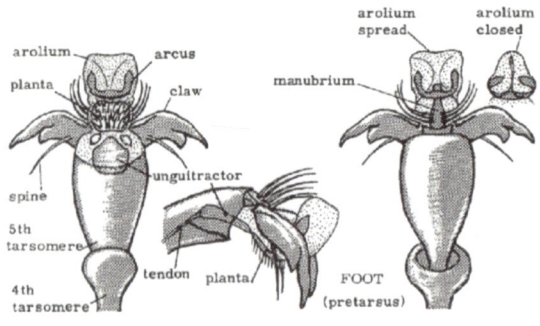

Fig 272. The foot is made up of several working parts known as the pretarsus, which is connected to the fifth tarsomere. Taken from Dade with kind permission of IBRA.

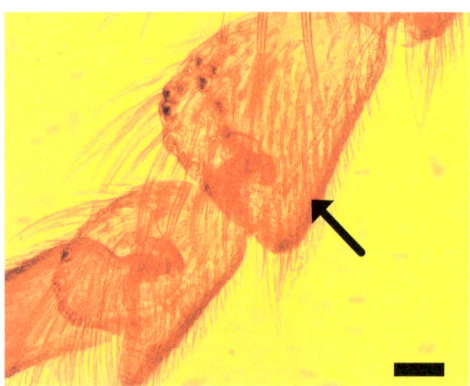

Fig 273. The unguitractor tendon passing through the tarsomeres. Scale bar 100 microns. X 100 Magnification.

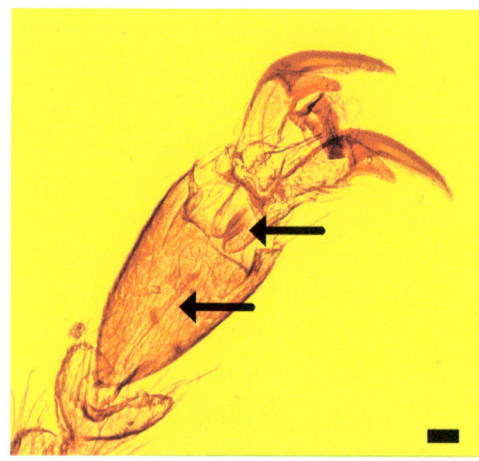

Fig 274. A.The unguitractor tendon in the pretarsus connecting to the unguitractor. Scale bar 100 microns. X 40 Magnification.

Each one of the claws is bi-lobed, consisting of a long curved tapering outer point and a smaller inner one; these are surrounded by stout spines. The claws are hinged via a tendon to the fifth tarsomere.

The claws of the worker and the queen are only slightly different in details (although the claws of the queen are much larger than those of the worker) but the drone's claws are larger and strikingly different in shape from those of either the worker or the queen, possibly to grip the queen during mating.

Fig 275. Worker's foot in the normal position with claws exposed, side view: Note the scythe-like shaped claw, showing the arolium, (lobe used for gripping on to smooth surfaces) in resting place on top and at the bottom the planta. X 40 Magnification.

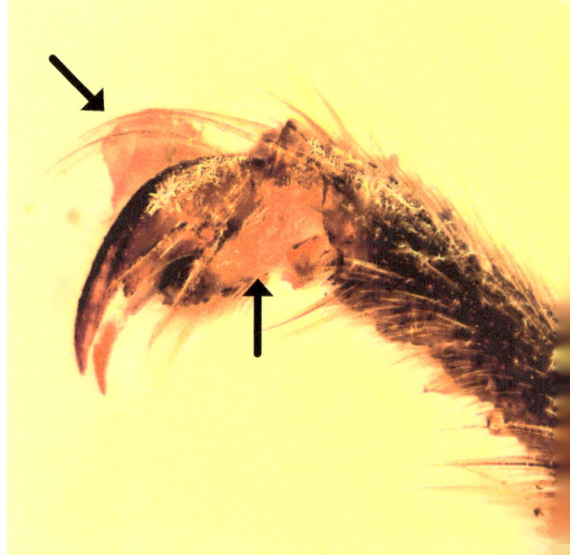

The arolium is a terminal lobe bent upward between the claws and is deeply cleft on its dorsal surface, and is made of a thick basal stalk

Fig 276. Queen's claw.
X 40 Magnification.

Fig 277. Drone's claw side view, much more angular in shape. X 40 Magnification.

whose walls contain several chitinous plates. It bears five long, thick, curved bristles projecting posteriorly over the terminal lobe. These act as touch sensors. The ventral side has numerous short thick spines.

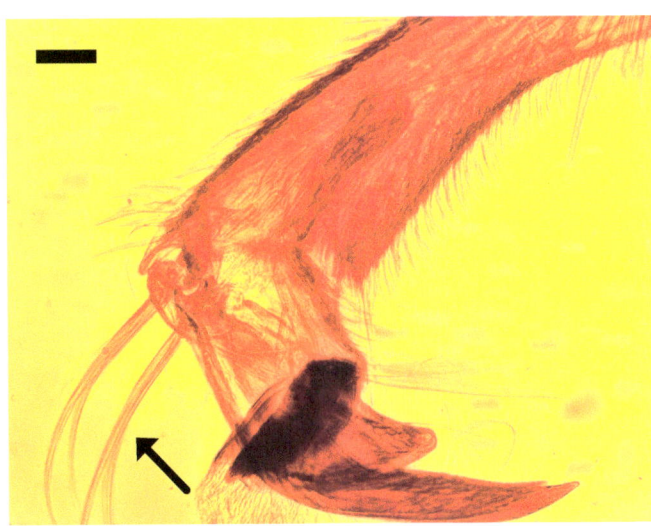

Fig 278. Long bristles that act as sensors. Scale bar 100 microns. X 100 Magnification.

When the bee walks on any ordinary surface it uses only its claws for maintaining a foothold, but when it finds itself on a smooth, slippery surface, like glass, the claw is of no use and the arolium is provided for such emergencies. The terminal lobe is pressed down against the smooth surface and its lateral halves are flattened, adhering by a sticky liquid excreted upon them by glands.

Around the inside base of the arolium is a u-shaped band, the arcus, which is joined at the lower end by a small sclerotized plate known as the planta, meaning' the sole of the foot,' which is covered in spines. This in turn is connected to another plate, the unguitractor, which has a tendon attached to it.

The arolium on the outside of the foot is connected by the manubrium meaning a handle; this is a plate bearing five or six long bristles.

Fig 279. Worker's foot outside front view, showing bi-lobe claws and arolium in a relaxed state on top and the manubrium. X 40 Magnification.

When the bee lands on a surface, the tendons are pulled by muscles in the femur and tibia; these act on the unguitractor plate which in turn draws on the membrane of the foot. The leverage on the claws flexes them downwards to grip onto the surface.

If the claws cannot grip then they continue to flex and spread sideways. This action is triggered by the long bristles on top of the foot, which act as sensors. The traction on the unguitractor plate is then transmitted to the planta plate, which acts on the arcus (an elastic u-shaped plate). This in turn pulls the arolium down and it spreads out to grip the smooth surface.

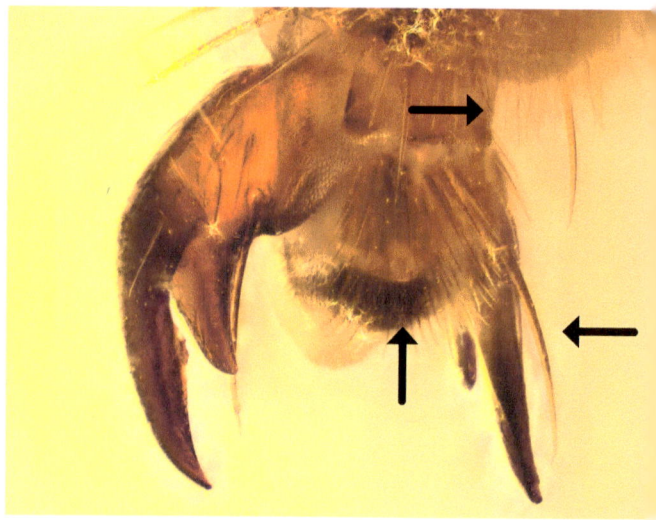

Fig 280 (above right). The underside of the foot showing spines right, arcus centre and unguitractor plate top. X 40 Magnification.
Fig 281 (right). SEM image of foot. © Dr Michel Asperges. Universiteit Hasselt Agoralaan gebouw D te BE 3590 Diepenbeek, Belgium. X 400 Magnification.

The bee's knees

Do bees have knees? The answer is they might have six.

In humans, the knee is the joint between the femur and the tibia. Since bees have a femur and a tibia in each leg, they could have knees. However, unlike us, their joints do not have a kneecap (patella) and thus technically they do not possess any knees.

According to the Oxford English Dictionary, the phase 'the bee's knees' phrase originated in the late eighteenth century meaning something very small. Its current meaning is believed to stem from American slang during 1920 referring to something outstanding or truly excellent. The origin might have come from the fact that the two words rhyme.

Some specialised features of importance to the bee are shown below.

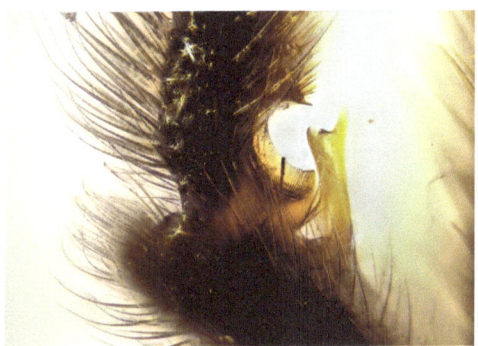

Fig 282. Worker antenna cleaner. X 100 Magnification.

Fig 283. Close up of queen's antenna cleaner. Scale bar 100 microns. X 100 magnification.

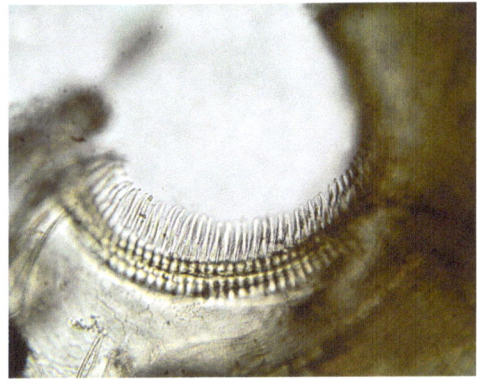

Fig 284. The antenna cleaner close up, showing hairs. © Dr Michel Asperges. Universiteit Hasselt Agoralaan gebouw D te BE 3590 Diepenbeek, Belgium.

Antenna cleaner in opened position on the foreleg, showing the notches in the comb with the fibula sticking up; this closes when the leg is bent to form a circular hole, allowing the antenna to be dragged through. The other hairs act as a brush to help clean the head of dust and pollen.

Fig 285. The pollen press on the hind leg showing the rastellum - large stiff like hairs on the top; these are used as a raking action on the opposite leg, collecting pollen from the hairs. The small hairs on the bottom are used to retain the pollen. Inside the bottom are the teeth on the auricle. The action of raking allows the pollen to be collected in the auricle and is mixed with saliva; the leg is then bent, squashing the pollen together and out onto the outside of the tibia, to be caught by the special curved hairs known as the corbicula. X 100 Magnification.

Fig 286. The single spine on the middle leg possibly used to collect pollen from the thorax; there is some conjecture over its usage or even if it is a leftover vestige. Scale bar 100 microns. X 100 Magnification.

Fig 287. Pollen pellet stored on the outer leg, having been pressed. Pollen grains collected on the inside leg, the basitarsus; these have pollen brushes in rows that are visible. X 40 Magnification.

Endocrine system

Endocrine (adjective.)

"Secreting internally," 1914, from endo- + Latinized form of Greek *krinein* "to separate, distinguish."

Endo-

Word-forming element meaning "inside, within, internal," from Greek *endon* "in, within."

Krei- crine

The proto-Indo-European root meaning "to sieve," thus "discriminate, distinguish."

It forms all or part of the endocrine; (n.2) "coarse sieve."

Fat bodies explained

A major and recent discovery is that the dreaded Varroa mite feeds on the honeybees' fat body and not on their haemolymph as previously thought.

There are three types of cells within the fat body:

Trophocytes form the major part of the fat body and are present in all stages of the bees life; Oenocytes (oil cells) are also found in all stages of larva, pupal and adult life, but the cells are destroyed at the larval stage and new ones formed in the adult stage. Large concentrations are formed over the wax glands in adult bees, peaking at the same time as wax production.

Fig 288A. Fat bodies, winter worker, the large white areas are air sacs. X 20 Magnification.

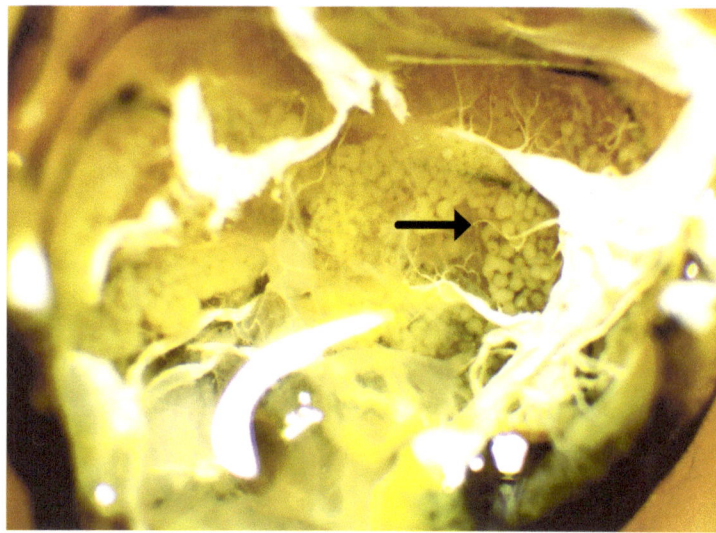

Fig 288B. Fat bodies, winter worker. X 40 Magnification.

Urate (excretory cells) are found only in the larva and pupa. They store nitrogenous waste in the form of uric acid and once the Malpighian tubules are formed, they disappear.

All insects have fat bodies; some are in the head but most are found just underneath the dorsal and ventral sinuses below the cuticle of the abdomen; these tissues contain lipids (fats), glycogen (sugar), triglycerides - which store chemical energy, and some albuminoidal granules (protein).

Fat bodies store and release energy according to the demands of the insect. The energy stored within these tissues is especially important during larval growth; when the cell is capped it amounts to 65% of the larval body weight.

In the pupal stage, fat bodies are released to build new cells. The cream colour of the larva is due to the density and whiteness of the fat tissue pressing against the transparent skin.

Another usage in adult bees is for storage when feeding is restricted. The summer workers have a very thin layer of fat bodies and can normally live for up to 38 days; their requirements are for high fat and low protein storage, mainly to convert as house bees into brood food. By comparison, winter workers have greatly enlarged fat bodies, which contain albuminoids; they act as reserves of food, enabling them to over-winter successfully for up to 140 days. Their requirement is low fat and high protein storage; they have few feeding or foraging duties to fulfil.

The metabolic action of fat bodies is often compared to the vertebrate liver, adipose tissues and bodily fat, as they store nutrients and synthesize proteins, lipids, and carbohydrates that circulate throughout their body. In another similarity to the liver, fat bodies help detoxify nitrogenous waste products, which they extract from the haemolymph- the blood, this process is unique to the insect phylum.

Interestingly, the workers live longer in queenless colonies due to the absence of brood rearing and demands on them to produce brood food from their hypopharyngeal glands and fat bodies, similar to the wintering workers.

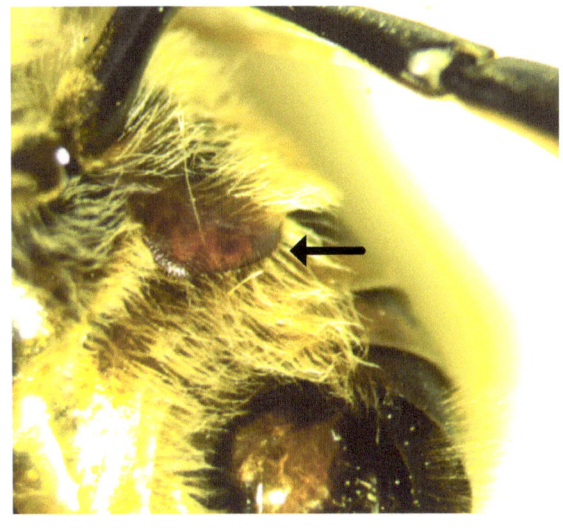

Fig 289. The underside of abdomen. X 30 Magnification.

The dreaded Varroa mite still attached to the bee, tucked up between the Sternites. This position is ideal as it offers a good location for piercing the skin where the fat bodies are situated immediately below.

Just in case you are wondering how the researcher discovered the method of Varroa mite feeding, he injected a phosphorus dye into the bees where it was absorbed by the fat bodies. The meal is then taken from the bee and the Varroa then reveals the glowing evidence!

Glands

The Thoracic Salivary Glands

The word 'Saliva' is derived from the early 15th century, from the French word 'alive', taken from the Latin word 'saliva' meaning spittle. The thoracic salivary glands are derived from the larval silk glands. They are found in the thorax and connect to the mouth. They are made up of bunches of short, round tubes. These are arranged on branches and connect to the main ducts. The secretions are stored in two small sacs that then pass into the median duct in the head. The queen, worker and drone all possess these glands. The enzymes produced by these glands aid digestion.

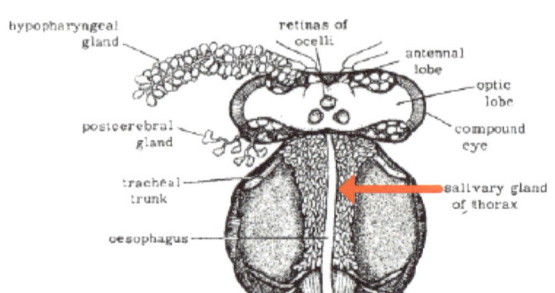 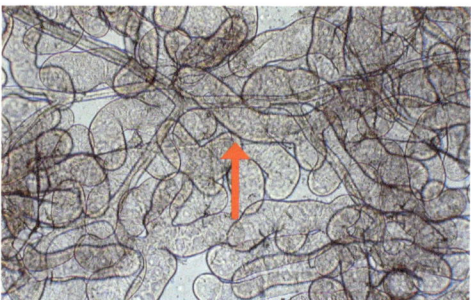

Fig 290A. Salivary gland of thorax, red arrow. Taken from Dade by kind permission of IBRA.

Fig 290B. Thoracic salivary glands, main ducts red arrow. © Dr Michel Asperges. Universiteit Hasselt Agoralaan gebouw D te BE 3590 Diepenbeek, Belgium.

The Mandible Glands

The mandible glands are found just above the mandibles on both sides of the head. In the worker bee, the saliva is stored in a lumen under pressure for use when required. From the lumen, a short duct enters into a groove on the inside of the mandibles. This aids in the feeding of the larvae. Both the queen and the drone have mandible glands.

The glands are single lobate, pod-like sacs, and are larger in the worker bee, rudimentary in the drone and of considerable size in the queen. Her secretions are commonly called queen substance. These secretions attract the drone in the mating flight. The workers' glands change when they start to forage, producing an alarm pheromone.

The secretions of these glands of both the drone and worker are translucent white, consisting of very small oily globules. Workers only produce brood food from these glands whereas drones use them to attract other

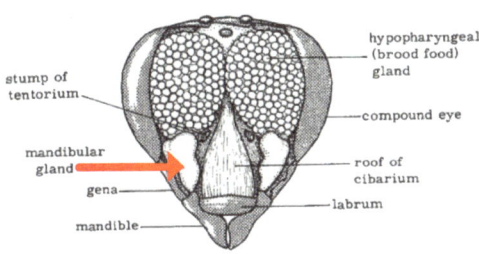

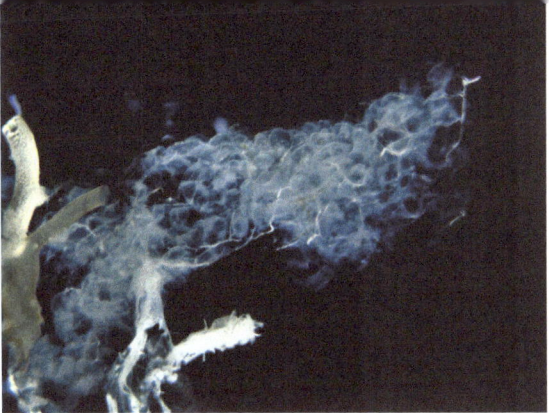

Fig 291A. Mandible gland, red arrow. The brood food gland shows a recently hatched worker bee that has not fed larvae yet. Taken from Dade by kind permission of IBRA.

Fig 291B. Darkfield illumination. Mandible gland. X 40 Magnification.

drones at congregation mating sites. The queen larva is fed almost 100% from the salivary glands of workers for the first three days of life. During the fourth and fifth days of life, a ratio of 1:1 mandible gland and hypopharyngeal gland secretions is given. These secretions consist mainly of protein made from both glands and are known as royal jelly.

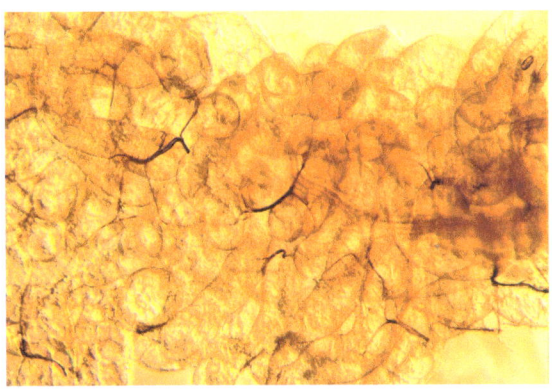

Fig 291C. Mandible gland enlarged. X 200 Magnification.

The Postcerebral Glands

These glands are found behind the brain and produce saliva. They join the ducts from the thoracic salivary glands in the base of the head, before entering the salivarium.

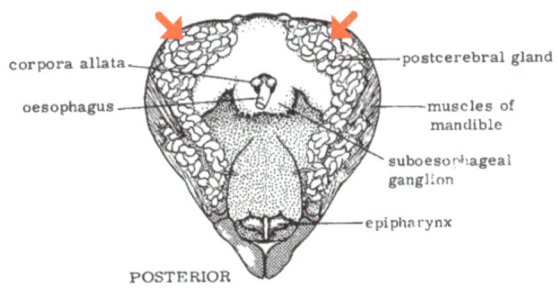

Fig 292A (above right). Posterior view of the postcerebral glands, red arrows. Taken from Dade by kind permission of IBRA.
Fig 292B (right). Dissected head of a worker bee, posterior view, showing the postcerebral glands, red arrows. X 40 Magnification.
Fig 292C (far right). The Postcerebral glands. X 200 Magnification.

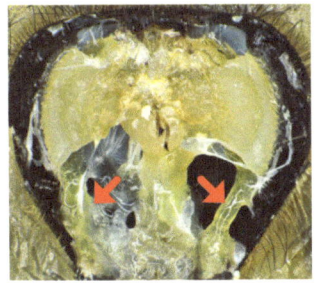
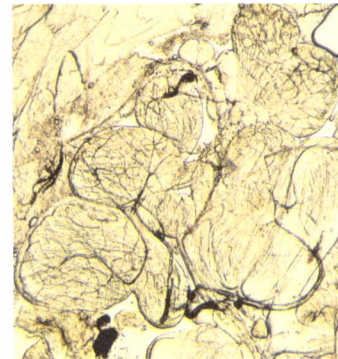

Hypopharyngeal Glands

The Hypopharyngeal glands or brood-food glands lie above the pharynx (the gullet). Their ducts run under the cibarium to the hypopharyngeal plate. Each gland is made up of a long, coiled tubule, which is connected by short side tubes, where several hundred small rounded bodies, the anci connect. These glandular tissues produce the fluid brood food that is fed to the larvae. They play no purpose in food digestion. In older worker bees that collect nectar, these glands shrink down and provide an enzyme that inverts the sucrose in nectar, which helps to produce glucose and fructose in honey.

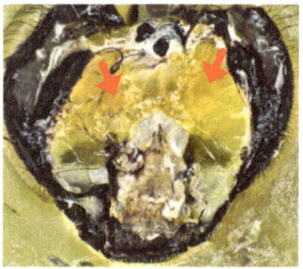

Fig 293A. Dissected head of a worker bee, anterior view, showing hypopharyngeal glands, red arrows.
X 40 Magnification.

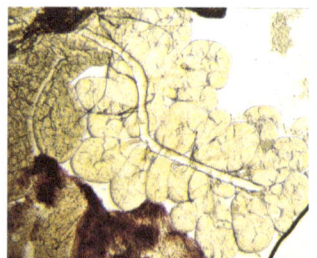

Fig 293B. The hypopharyngeal gland, showing the tube and the anci.
X40 Magnification.

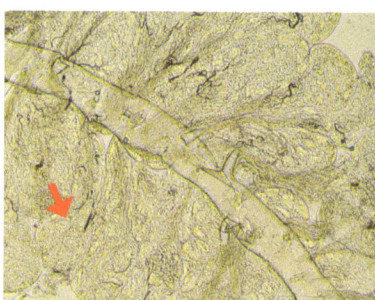

Fig 293C. The hypopharyngeal gland, showing the small branching tubes and the anci, red arrow.
X200 Magnification.

The Mouth

The opening of the mouth has the labrum above and two mandibles at either side, with the proboscis below. The mouth has two salivary ducts that enter the bottom area named the salivarium, which acts as an ejection pump for saliva. Just above this, the cibarium or food chamber is situated. On the base of the cibarium lies the hardened hypopharyngeal plate. The front lobes bend downwards. Two long arms run backwards under the cibarium acting as support and for muscle attachment. A group of 50-60 hypopharyngeal sensillae are found on either side of the entrance to the cibarium, these are believed to act as chemoreceptors, sampling material in and out of the cibarium.

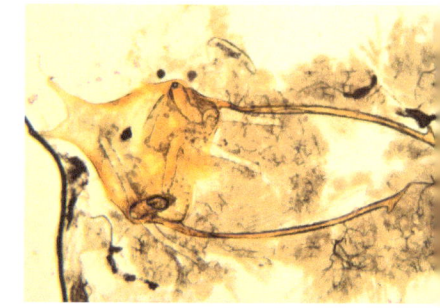

Fig 294A. The hypopharyngeal plate of a worker bee.
X 40 Magnification.

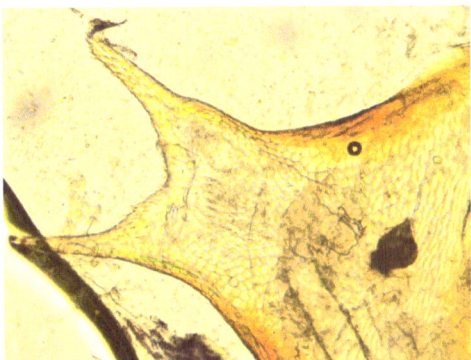

Fig 294B. The hypopharyngeal lobe of a worker bee. X 200 Magnification.

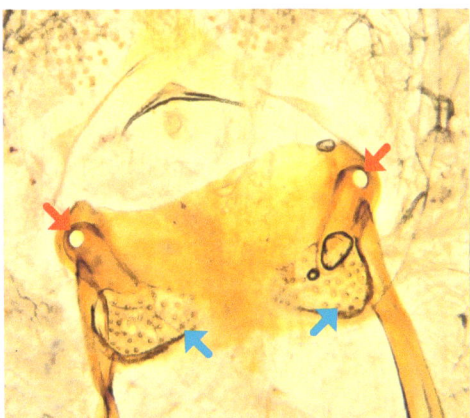

Fig 294C. The hypopharyngeal plate of a worker bee, showing where the ducts enter into the mouth, red arrows. Salivary ducts orifices, blue arrows. X 100 Magnification.

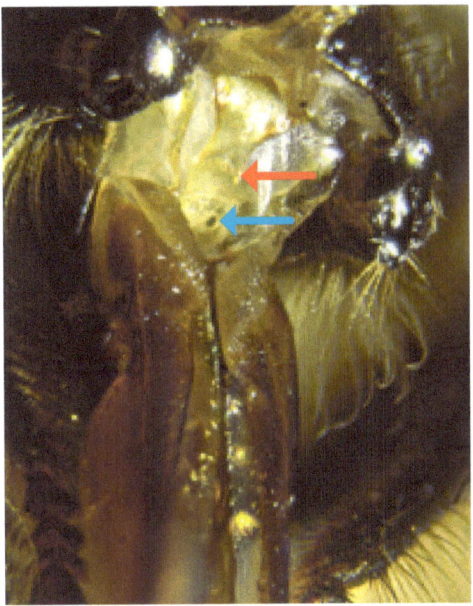

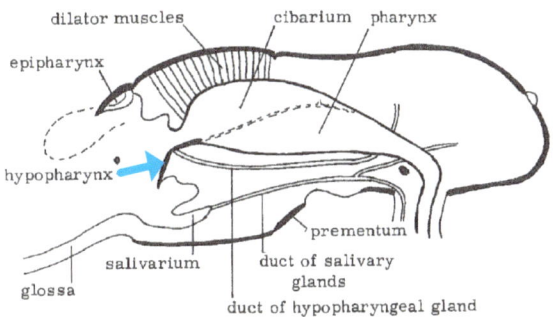

Fig 294D (above). The mouth looking into the salivarium area, blue arrow. The hypopharyngeal lobe of a worker bee, red arrow. X 40 Magnification.
Fig 294E (above right). The cavity of the mouth and associated structures. Blue arrow hypopharynx lobe. A diagrammatical, longitudinal section through the head. Taken from Dade by kind permission of IBRA.
Fig 294F (right). The mouth looking into the salivarium area. The red arrow showing small sense hairs. X 100 Magnification.

Fig 294G. A view into the opened mouth, the entrance to the cibarium, the red arrows showing the entry canals of the hypopharyngeal glands. The mandibles have been removed for clarity. The blue arrow, showing the epipharynx. The white arrow, showing a group of 50-60 hypopharyngeal sensillae. X 40 Magnification.

Nasonov gland

Workers have a scent gland named after the Russian zoologist who discovered it, Nikolai Nasonov. It is located at the tip of the upper abdomen, between the sixth and seventh tergites. The gland emits a mixture of seven terpenoids, which serve primarily as an attractant. To release the chemical mixture the workers bow their heads with their abdomen elevated and tilt the last abdominal segment downward while fanning the wings to disperse the pheromone. Bees use this scent to help other workers of their colony to locate their hive, food, and water sources. It also acts with queen substance in a pheromone concert to keep the bees of the swarm together. Known as semiochemicals.

The glands are normally concealed beneath the overlapping tergites. The drones do not possess this gland.

Fig 295. Typical scenting pose.

When the worker exposes her intersegmental membrane it pulls it flat showing a ridge, the ante-costa, (before the coast) beneath this is a collection of between five and six hundred glandular cells that secrete a liquid via small ducts to the outer surface through pores. The released pheromone is collected into a channel just above the ridge.

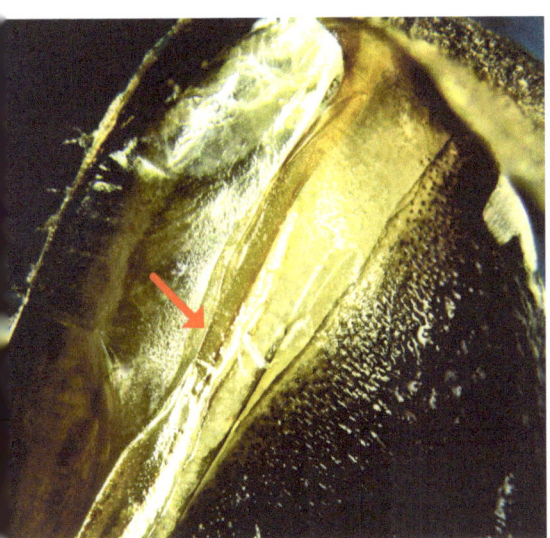

Fig 296. View of the intersegmental membrane-showing ridge. X 40 Magnification.

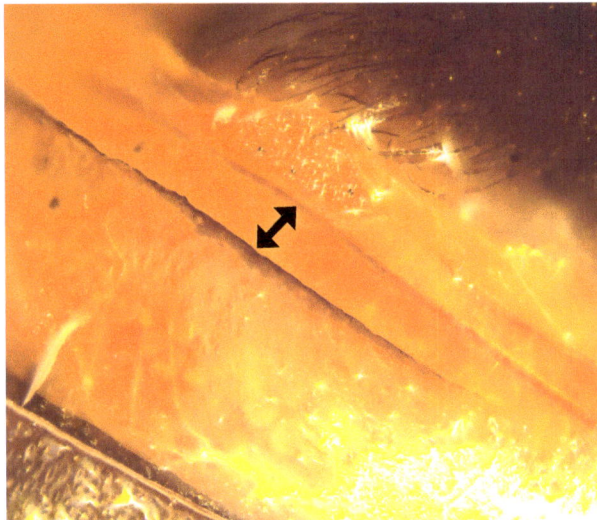

Fig 297. Close up view of ridge, showing antecostal line and canal. X 100 Magnification.

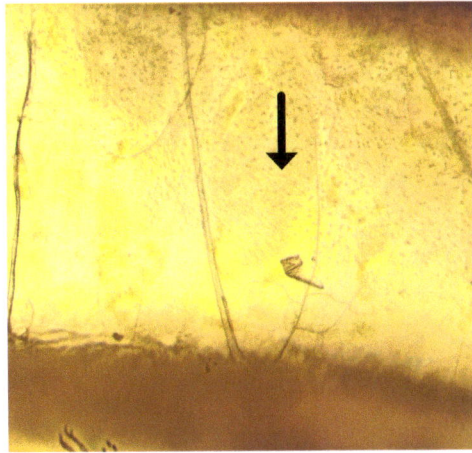

Fig 298. Close up view of membrane showing pores. X 200 Magnification.

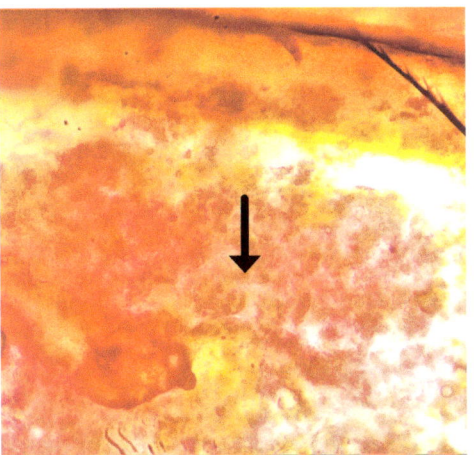

Fig 299. Stained gland cells under the membrane. X 200 Magnification.

Waxing lyrical

The construction of a wax cell starts with worker bees holding on to each other's legs, closely together at a temperature of 35-37 degrees centigrade; this heat is provided by the bees. They secrete wax drops from the four pairs of wax glands found in between the last four abdominal sternite segments. The worker bees use their hind legs that are equipped with specialised hairy bristles, to collect them and then to move the wax plates forward to their head, where they then manipulate them using their mandibles into round shapes on the comb. When they touch the other warmed circular cells they form a hexagonal shape. The brace comb and edge comb will be irregular in shape and size.

Very young and old bees cannot secrete wax. Under normal hive conditions, they produce wax from between the twelfth and eighteenth days of their life. It is estimated that an average hive contains about 1.2 kilos of wax. There is evidence that bees will orientate the comb in certain directions when a new colony forms, by either using their memory and build the comb in the direction of their old colony or will use the earth's magnetic field for orientation.

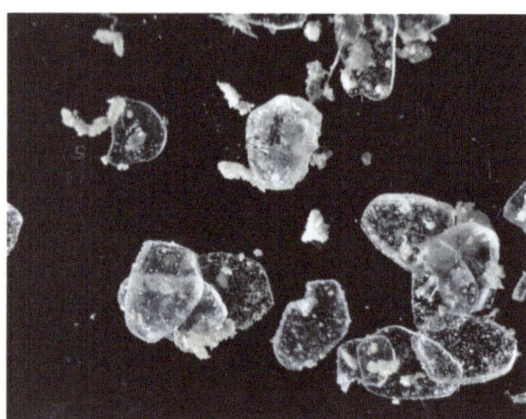

Fig 301. Darkfield magnification of wax plates, taken from hive floor. X 10 Magnification.

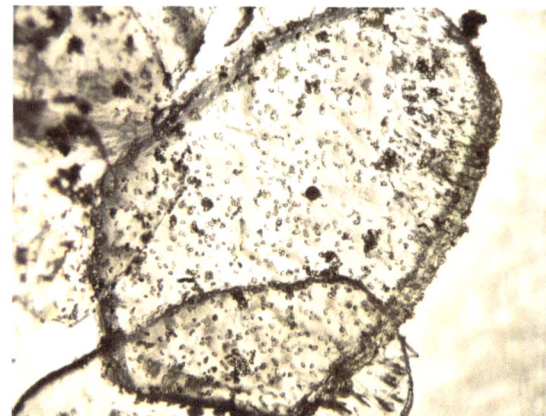

Fig 302. Bright field magnification of wax plates, pollen grains coating outside, collected from hive floor. X 40 Magnification.

There are two sizes of cells, the first around 5.4 mm for the development of worker bee, the second one measuring 6.4 mm for the much larger drones. Drones that are reared in worker-sized cells develop normally but are much smaller in size. Smaller cells at the top of a frame, the same width as worker cells but often up to twice the depth, act as storage for food and are replenished regularly by house bees; excess is stored in supers, above the brood area. The capping of food cells is white as it is made from fresh

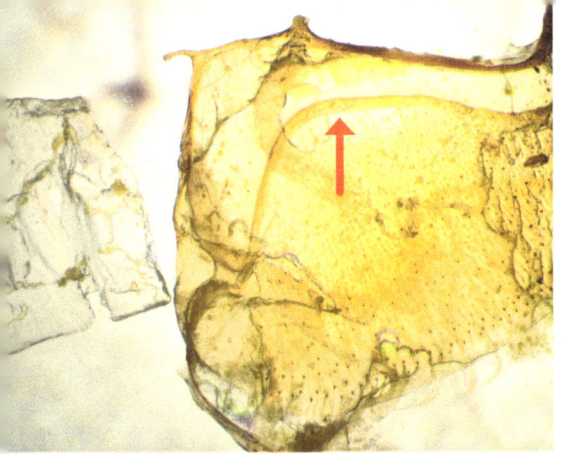

Fig 303. Internal view of the wax mirror, note wax plate curved shape.

Fig 304. Wax plates between sternites. Photograph by Helga Hilmann and is from the wonderful book "The Buzz about Bees" by Jurgen Tautz.

wax. However, the colour will change over time as the outer wax coating of pollen stains them yellow. The wax of the breeding cells also gets darker with use, created by the shedding of the outer covering of the pupae and voidance of their waste matter once the cell is capped. The walls of the cells become thicker over each breeding cycle and the internal cell size is reduced slightly, to around 5 mm, producing smaller workers. This is due to the residue left behind and new wax laid by the cleaner bees. The darker comb also attracts wax moths as it contains more protein residue. A dark comb is also a good attractant for collecting a swarm.

Fig 305. An older drone feeding itself with honey. Note the colour of the capping.

The capping of brood cells consists of silk from the pupal cocoon, pollen, new white wax and propolis, giving them a more yellow appearance. Drone capping is more domed in shape. These cappings are reused by the house bees.

It has been observed that up to 15% of the comb in a feral colony will be drone cells. Most beekeepers insert wax frames with smaller worker patterns on to stimulate wax building, which appears to reduce the amount of drone comb to 5 %. Workers will build drone comb only in the spring before the swarming season starts; it appears there is feedback to the workers as to the amount of drone comb already present, thereby limiting any further construction.

The hexagonal shape allows for the maximising use of space and wax usage. The cells also tilt backwards between nine and fourteen degrees.

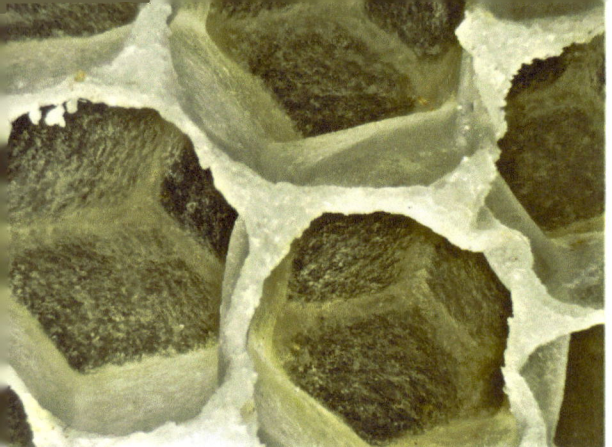

Fig 306. Worker comb cells - note the opposite side is offset.

Fig 307. Capped cell cut open to show pollen - note different colours from the different flowering source.

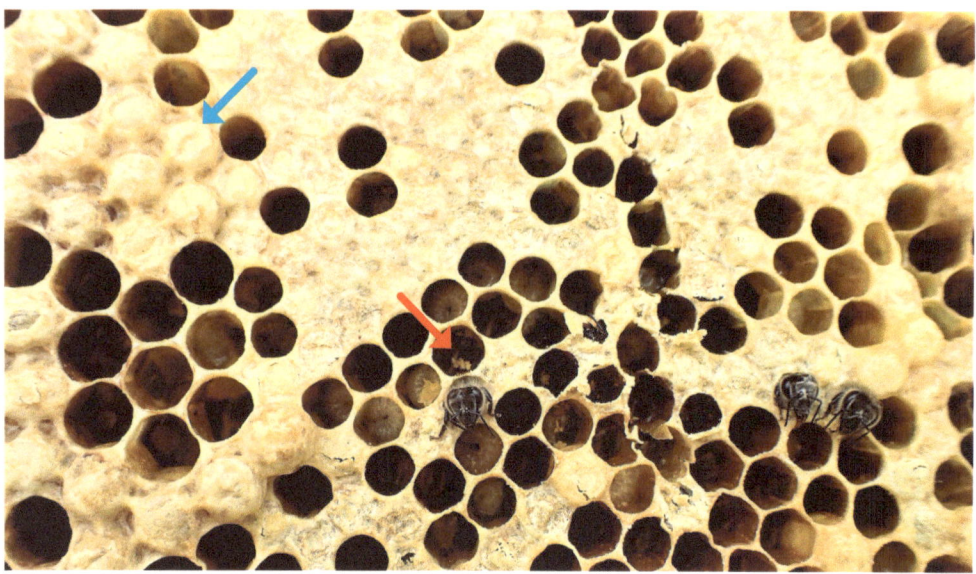

Fig 308. Mixed cells, workers emerging from cells - larvae red arrow; raised drone comb - blue arrow.

Fig 309. Section through capped worker larvae, head on the right.

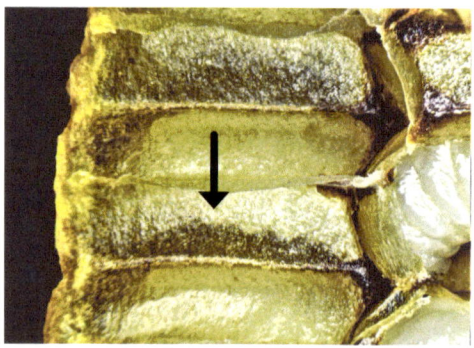

Fig 310. Old cells, showing staining from old pupal coating and excrement.

Martyr to the cause

The drone and worker bees have different roles to play within the hive. As its name denotes it is the worker who is the industrious one; the other is built for reproduction, passing the queen's genes onto the next generation. They both share an altruistic character, meaning that they will sacrifice themselves for the greater good of the colony. The drone dies after copulation with a virgin queen as his endophallus is left behind in the queen's sting chamber as a mating sign.

The worker, and only the worker, will defend the hive by stinging her foe. The drone lacks this defence mechanism. The sting cannot be withdrawn once it has been used on mammalian skin, owing to the toothed barbed design at the end, which locks it into place. Because of this locking method in mammalian skin, the stinging mechanism as a whole is torn out and the worker dies. The remaining sting body releases a pheromone to attract other workers to help defend the colony. The seventh nerve ganglion is also torn out; this controls the muscles which keep pumping venom into the wound. A freshly collected sting from a bee suit can successfully be viewed under a microscope to witness this pumping action.

 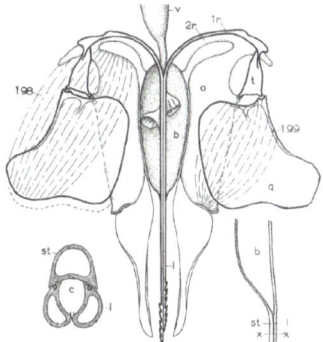

Fig 311 (left). The sting mechanism. v. duct of venom gland. b. bulb. l. lancet. o. oblong plate. Q. quadrate plate. t. triangular plate. 1r. first ramus. 2r. second ramus. st. stylet. c. venom canal. Taken from Dade with kind permission of IBRA;

Fig 312 (right). Flattened dissected sting mechanism. X 20 Magnification.

The sting is situated within the last segment of the abdomen - it is often illustrated flat but is a round shape when viewed in real life. The sting resides in a soft flexible sheath that is fixed to the roof cavity of the abdomen when not in use. There are three parts to the sting - the two lancets, which are keyed into the third part, the stylet body, located behind them, enabling them to glide together. The abdomen is tightly curled up until the lances protrude and the act of stinging commences.

At the base of the stylet are two ducts that link into the poison gland; they have a non-return valve and send poison into the bulb and then into the channel formed by the lances and stylet.

To sting the large protractor muscles attached to the quadrate plate and triangular plate flex against the oblong plate, which is fixed and does not move. This enables the sting mechanism to swing outwards and forwards, with the aid of the curved rods, the ramus. The second set of muscles at the rear of the oblong and the quadrate plate contract backwards. This action drives the lancets at each protraction and retraction, allowing them to move slightly forward and deeper rapidly. The queen has more curved lancets and the basal apparatus are different in size and shape. She uses her sting only within the colony to kill rival queens.

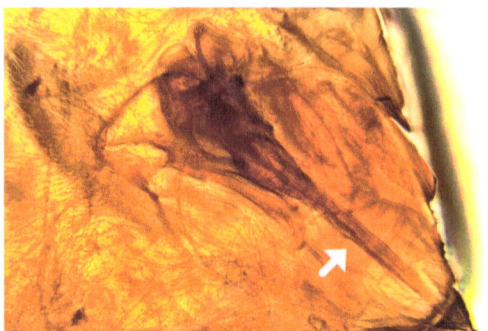
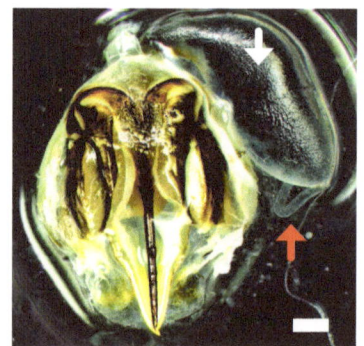

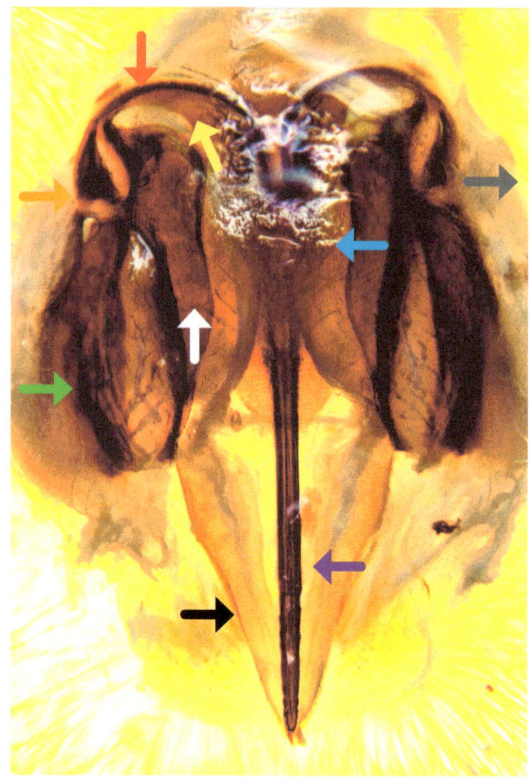

Fig 313 (above left). The barbs on the lances. Scale bar 0.5 mm. X 100 Magnification.
Fig 314 (above middle). Cleared specimen, transmitted light showing sting mechanism. X 20 Magnification.
Fig 315 (above right). Taken from a bee suit. A sting mechanism and venom sac, white arrow. Red arrow, acid gland duct. Scale bar 200 microns. X 10 Magnification.
Fig 316 (right). Transmitted light showing the front of sting mechanism in the true position. Black arrow, sheath. Purple arrow, lancets. Blue arrow, bulb. Yellow arrow, second ramus. Red arrow, first ramus. White arrow, oblong plate. Orange arrow, triangular plate. Grey arrow, muscle. Green arrow, quadrate plate. X 40 Magnification.

The Sensory Organs

Sensory (adjective.)

1749, "Pertaining to sense or sensation," from Latin *sensorius*, from *sensus*, past participle of *sentire* "to perceive, feel".

Organ (noun.)

The biological meaning "body part of a human or animal adapted to a certain function" is attested from late 14c, from a Medieval Latin sense of Latin *organum*.

Antennae, horizontal masts

Antennae are movable sensory organs located on the head of most arthropods, normally in pairs although spiders and Varroa have none. Antennae are segmented, usually located above or between the eyes and serve different sensory functions for different insects. In general, the antennae might be used to detect odours and tastes, wind speed and direction, heat and moisture, and even touch. In some insects, they may even serve a non-sensory function, such as grasping prey. Because antennae serve different functions, their forms vary greatly-there are about thirteen different antennae shapes and their form may be an important key to insect identification. Those found mainly in ants or bees are called geniculate antennae, from the Latin for knee, as they are bent or hinged sharply.

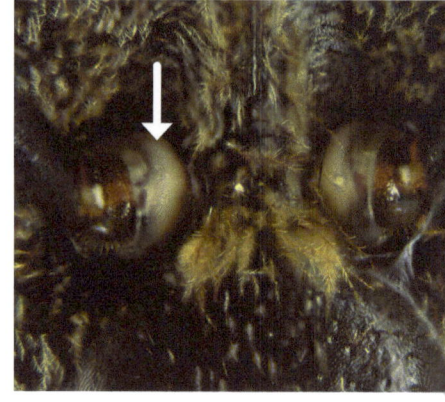

Fig 317. Basel knob of antennae, a ball and socket type joint.

To navigate their scent-filled environment insects require a sophisticated system of odour detection that is better than humans.

They do not have noses the way mammals do but use their antennae or sense organs in their legs or feet to detect chemicals in the air. An insect's acute sense of smell enables it to find mates, locate food, avoid predators and even gather in groups such as the drone congregation area.

Some insects rely on chemical cues to find their way to and from a nest, think of bees and wasps.

Insects produce semiochemicals, or odour signals, to interact with one another. Most of the olfactory sensilla, or smell-gathering organs, are in the insect's antennae; in some species, additional sensilla may be located on the mouthparts, feet or even the genitalia.

Scent molecules arrive at the sensilla and enter through a pore.

However, simply collecting the chemical cues is not enough to direct an insect's behaviour. This takes some intervention from the nervous system.

Special cells within the structure of the sensilla produce odour-binding proteins which capture the chemical molecules and transport them through the lymph to a dendrite, an extension of the neuron cell body. Odour molecules would dissolve within the lymph cavity of the sensilla without the protection of these protein binders. The odour-binding protein now passes its companion smell to the receptor molecule on the dendrite's membrane. This is where the magic happens. The interaction between the chemical molecule and its receptor causes a depolarization of the nerve cell's membrane which triggers a neural impulse that travels through the nervous system to the insect brain, informing its next move. The insects smell the odour and will pursue a mate, find a source of food, or make their way home, accordingly.

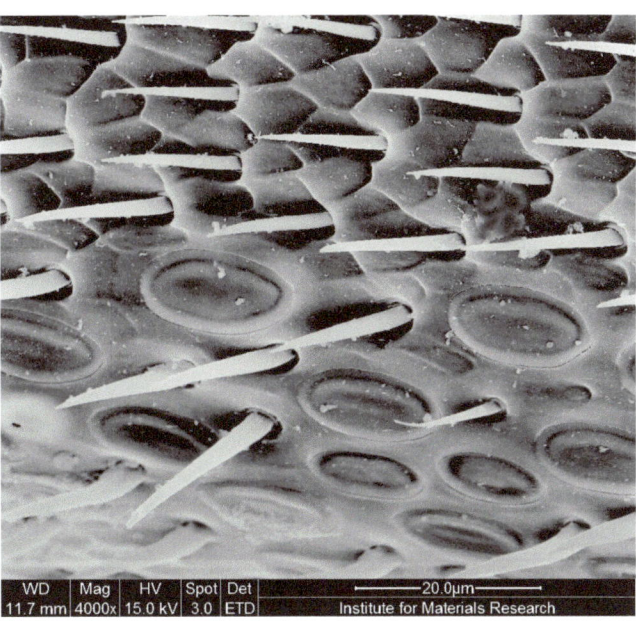

Fig 318 (right). SEM Sense plates and sense hairs on the antenna. © Dr Michel Asperges. Universiteit Hasselt Agoralaan gebouw D te BE 3590 Diepenbeek, Belgium X 4000 Magnification.

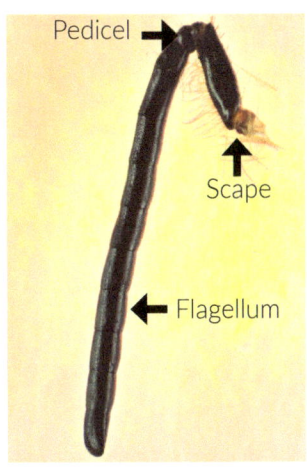

Fig 319 (above). The drone's antenna. X 10 Magnification.

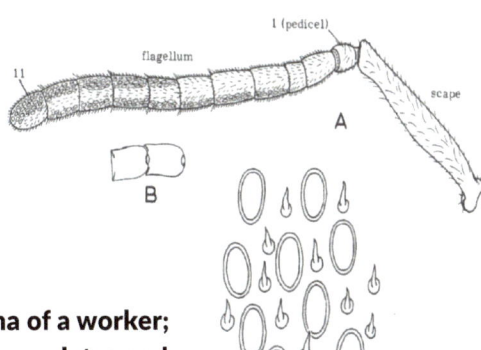

Fig 320 (right). The antenna of a worker; B. jointing segments; C, sense plates and sense hairs on the antenna. Taken from Dade with kind permission of IBRA.

The scape has two muscles internally to rotate the antennae on a ball and the socket joint on the head. The flagellum and pedicel, by comparison, are kept ridged by blood pressure. Inside the pedicel is the Organ of Johnston.

The word antenna (plural antennae) comes from the Latin antemna, meaning a 'horizontal mast spar' that was designed to spread a square-rigged sail. Among other uses, the term is used for the feeler organ on the head of an insect, crab or other creature. The bees' world is governed by cuticular hydrocarbons and pheromones, which give them information as to the state of the colony, so smell is a very important means of information and communication inside and outside the hive. Pheromones are substances that are secreted to the outside of the body by exocrine glands of which the bee has several. These are picked up by the other bees (and sometimes-other insects) and depending on what gland it came from, a specific reaction can result - behavioural, developmental or physiological. Examples of this might be the alarm pheromones given off when stinging, queen substance uniting the colony and the Nasonov pheromone produced from a gland on the top of the abdomen of workers and used at the entrance to indicate the hive location to returning foraging family members. All bees can smell through their antennae and incidentally, feet.

The antennae have four segments: the ball and socket joint connecting with the head, the longer scape (an upright stem), and then the small, bent top part called the pedicel (small stalk), followed by the long flagellum (little whip) - the sensory part of the antenna. These have 10 annuli in the worker and eleven in the drone, on which are situated various hairs and specialised pegs and pits that have porous areas, which allow odours to permeate.

It is within the pedicel that the collection of sensory cells known as Johnston's organ is situated which is used for determining airspeed when in flight and sound detection.

Drone antennae have about double the surface area of a worker bee, containing approximately 16,000 plate receptors (worker bees have 2,700) which are thought to be used in seeking out the queen pheromone. It has been found that drones can smell the queen pheromone up to 800 metres away when flying. The worker bee scores higher on taste receptors on the antennae - 2,000 against 400 for the drone - reflecting the unique role they both have in supporting the colony.

The scape has two muscles internally to rotate the antennae about on its ball and socket joint on the head; the flagellum and pedicel are kept rigid by blood pressure only.

These (meaning the scape, flagellum and pedicel) have pitted oval pores and small peg-like hairs on the outside. The important ones are found primarily on the eighth annuli of the flagellum. The sensilla picola are specialised individual oval sense organs connected to nerves, which allow the drone and worker to taste. The smell is sensed by the Basiconic and Trichoid sensilla, the hairy peg-like structures; other sensors act as strain gauges, sensing stress and movement such as how far the head will turn. Note that carbon dioxide levels, wind speed and temperature are sensed by other sensillae both internal and external.

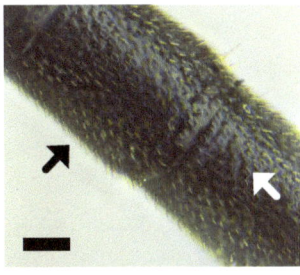

Fig 321. Sensilla on the antennae. White arrow, sensilla. Black arrow, peg-like hairs. Scale bar 100 microns. X 40 Magnification.

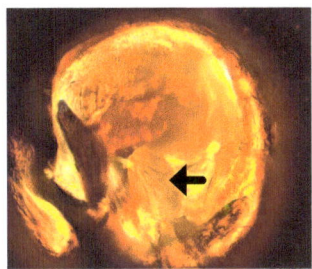

Fig 322. Internal view of scape, showing muscles. X 100 Magnification.

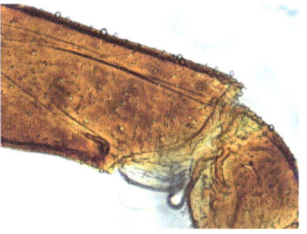

Fig 323. Cleared specimen showing pedicel and scape joint. X 40 Magnification.

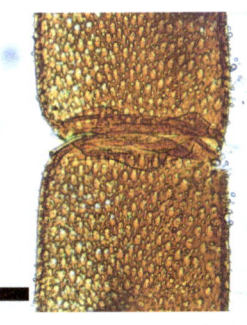

Fig 324. Flagellum joint. Scale bar 100 microns. X 40 Magnification.

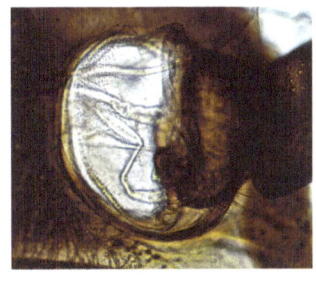

Fig 325. Basel knob of the ball and socket joint showing muscles entering the scape. X 40 Magnification.

Fig 326. Dissected head, showing the four muscles inside, that move the scape, red arrows. Antenna, blue arrow. X 40 Magnification.

Fig 327. A dorsal hair plate. Scale bar 100 microns. X 40 Magnification.

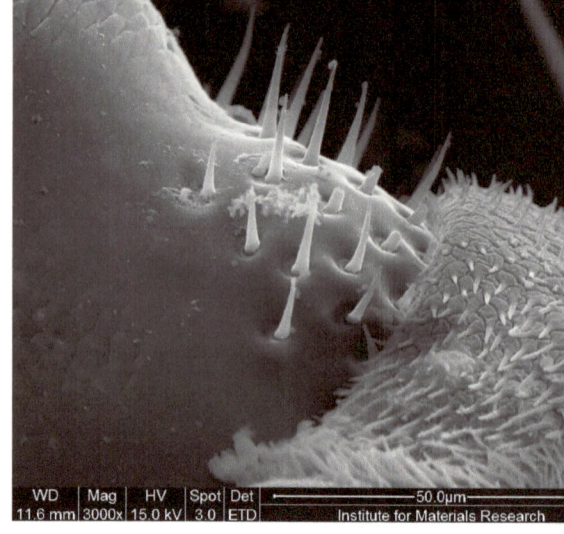

Fig 328. SEM image of dorsal hair plates. © Dr Michel Asperges. Universiteit Hasselt Agoralaan gebouw D te BE 3590 Diepenbeek, Belgium. X 3000 Magnification.

The dorsal hair plate is found at the hinge joint between the pedicel and the scape. This joint allows the pedicel and flagellum together to move up and down relative to the scape. There are three sets of plates, one on top and two on each side of the pedicel. The short spines act as mechanoreceptors, allowing the antennae to move giving feedback when it deflects the hairs. They are asymmetrical in shape, being smaller in some areas, which indicate a preferred area of movement.

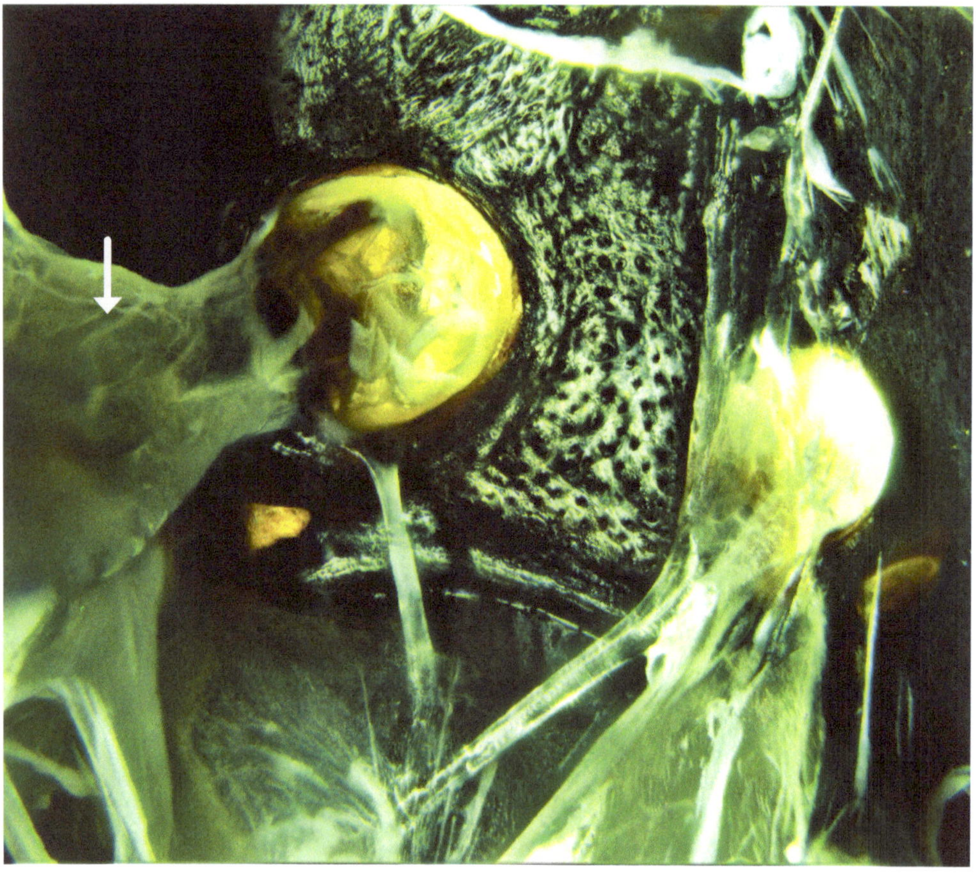

Fig 329. Dissected head showing internal muscles attached to the antennae. X 100 Magnification.

Organs of Jacobson and Johnston, smell and the hearing bee

Because we breathe through our noses scent molecules are carried onto a receptive area inside the nose; from where the chemical message is passed to two swellings on either side of the olfactory bulb. Known as the 'Organ of Jacobson', they are named after the sharp-eyed Danish anatomist who first discovered them.

Dogs are reported to have up to 1000 times more sensitive to odours than we do.

Fish have sensors that sense smell situated on their bodies, eels and salmon find their way back home halfway across the world to mate, using the chemical odours in the water.

The male moths have massively feathered antennae to search out female pheromones during the breeding season.

Moths are known to hear the echo given out by bats, because of this they literally drop down out of the way from this predator's grasp. The array of sensing methods varies for each species and although most have the Organ of Jacobson, it is used in different ways.

What about the honeybees' sense of smell and hearing?

Insects have a similar system discovered by a physician named Christopher Johnston. Other insects, unlike honey bees, lack chemoreceptors which are used for the detection of movement, wind speed and vibration.

Inside the antennae, at one end of the Pedicel and the Flagellum, a flexible membrane is suspended; it is here that the vibration is picked up by the Organ of Johnson. This is used to tell the bee the airspeed and for hearing sound.

Bees generate sound not only through the movement of their wings but also with their thoracic muscles. Although they use these muscles to move their wings, they can use their wings to produce heat and generate acoustic signals. Travelling sound waves have both pressure and particle movement components. Sound waves are measured by their frequency in Hertz (Hz), or cycles per second. The frequency of sound waves is heard as pitch; a higher wave frequency creates a higher pitch. Honeybees produce many frequencies of vibration and sound – from less than 10 to more than 1000 Hz. So far, it has been shown that they can detect sound frequencies up to about 500 Hz. The human hearing works with a tympanic membrane vibrating inside the ear; when young, we hear between 20 -20,000 Hz.

200–300 Hz is the sound range generated during the waggle dance, with about 15 bursts of sound every second. Attending bees have vibration sensors on their legs, named the subgenual organs (meaning below the knee) located in the tibia near to the joint with the femur, which act to pick up the vibrations through the comb. The bees then press their thorax against the comb and vibrate their wing muscles at a frequency of 350 Hz. When the dancer perceives these signals it may stop dancing and deliver small samples of her collected food to the attendants.

Another sound heard in the colony at times is the queen piping, which occurs more commonly when there is more than one queen in a hive; it is believed that the piping is a signal that a virgin is ready to fight for the honour of being the one-and-only. During swarm season, workers hearing the sound may try to keep the virgins separate to have more than one queen available in case she is needed. Queen piping ranges from 200-500 Hz.

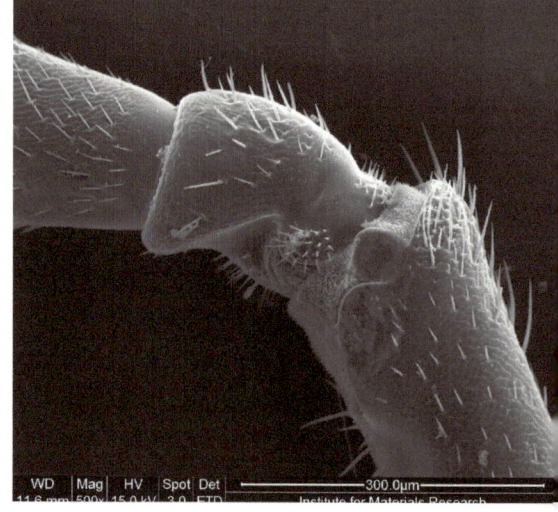

Fig 330 (above right). Pedicel x 40 magnification. X 200 Magnification.

Fig 331 (right). SEM image of Pedicel. © Dr Michel Asperges. Universiteit Hasselt Agoralaan gebouw D te BE 3590 Diepenbeek, Belgium. X 500 Magnification.

Photon collectors

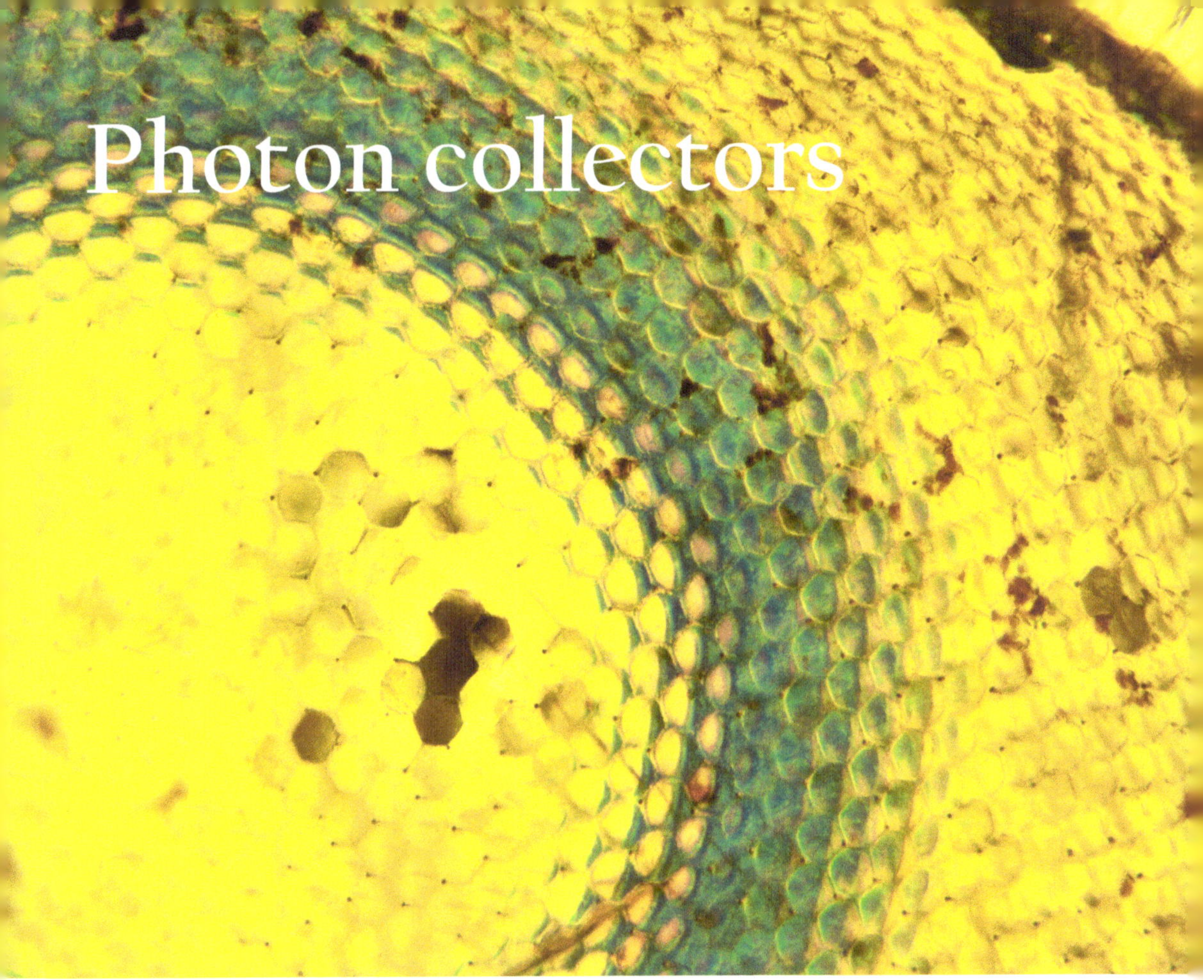

Fig 332. The dissected lens looking at a circular blue image over the transmitted light. Note how the facets display the image outwards. This represents a close-up view, not how the bee will see from a distance or flying. X 40 Magnification.

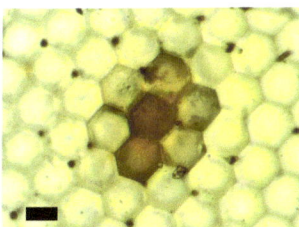

Fig 333. Internal view of hexagonal facets. A group of these facets can make up a 'region of vision'. X 200 Magnification. Scale bar 20 microns.

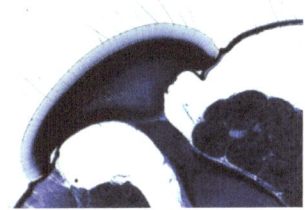

Fig 334. Section through compound eye. © Dr Michel Asperges. Universiteit Hasselt Agoralaan gebouw D te BE 3590 Diepenbeek, Belgium

Fig 335. The rear of the head with chitin removed to show retina. The Blue arrow shows the lamina and the red arrow the medulla, the hindmost section of the brain. X 100 Magnification.

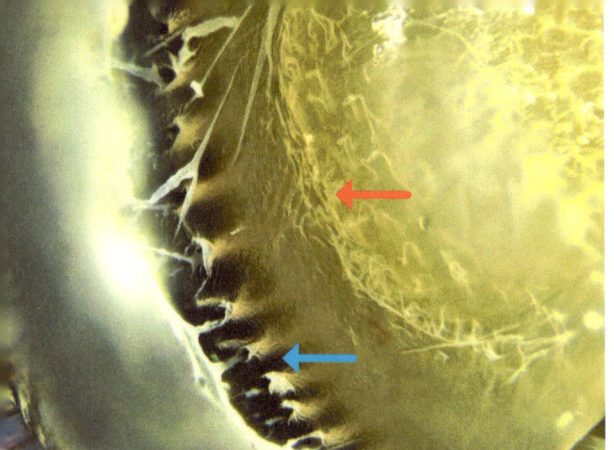
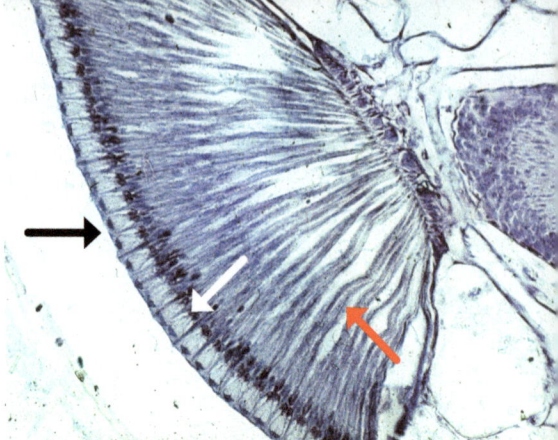

Fig 336. Section through the head, showing retinal nerves blue arrow and the lamina of the optical lobe red arrow. X 200 Magnification.

Fig 337. Enlarged section through the eye. Lens, black arrow. Crystalline cone, white arrow. Ommatidia the light-sensitive pigment cells, red arrow. © Dr Michel Asperges. Universiteit Hasselt Agoralaan gebouw D te BE 3590 Diepenbeek, Belgium

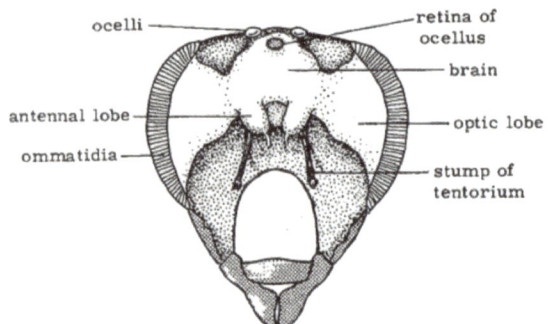

Fig 338 (left). Section through head showing brain and optic lobe and ommatidia. Taken from Dade by kind permission of IBRA.

Externally the eye is made up of thousands of individual ommatidium (Greek for 'little eye.') The top layer consists of hexagonal lenses that measure about 0.02 mm in diameter and 2.6 square mm. The worker has about five thousand, the queen about three thousand. The drone has up to eight thousand and they are larger, measure 0.03 mm in diameter and 9 square mm, reflecting the role his eyesight plays in the mating process.

The ommatidium is a long tapering tube connected to the optic lobe. Each provides a separate angular field of view. They consist of a clear convex lens, then a transparent crystalline cone that is surrounded by pigment cells to stop light from interfering with the next cone. Connecting the cone is a long transparent rod called the rhabdom (Greek for rod) which is surrounded by eight retinula cells. Some are sensitive to blue light, others to green light and one is sensitive to ultraviolet light. The ommatidia connect at the other end via nerves to the optic lobes. The optic lobes are made up of nerve fibres and nerve cells composed of three distinct parts - the outermost area is the lamina ganglioaris, then medulla externa and finally the medulla interna.

Bees' eyes detect cues (things that serves as signs) and direction of movement in each 'region' of the eye. Drones use several regions (a group of lenses acting together) to help detect the movement of the queen from afar. This is known as retinotopy (Greek for a place) and allows for the mapping of visual input from the retina to neurons. They use their sight to detect the edges only of a shape and look at areas separately; they have no concept of texture, shape or topology.

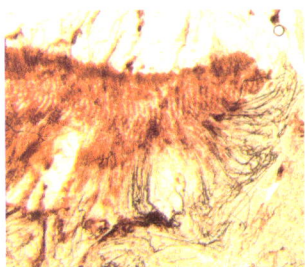

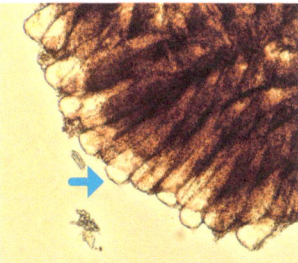

Fig a. Slide of the ommatidium showing the long tapering tubes surrounded by pigmented cells. Taken from a fresh specimen. X 10 Magnification.

Fig b. A stained slide is taken from a bee pickled in alcohol. The arrow shows the transparent crystalline cones. X 40 Magnification.

Fig c. Stained slide is taken from a bee pickled in alcohol. The arrow shows the transparent crystalline cones. X 200 Magnification.

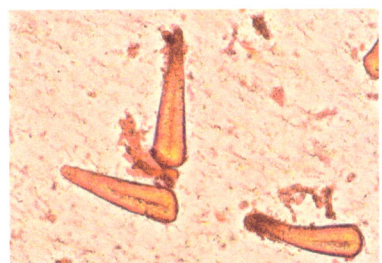
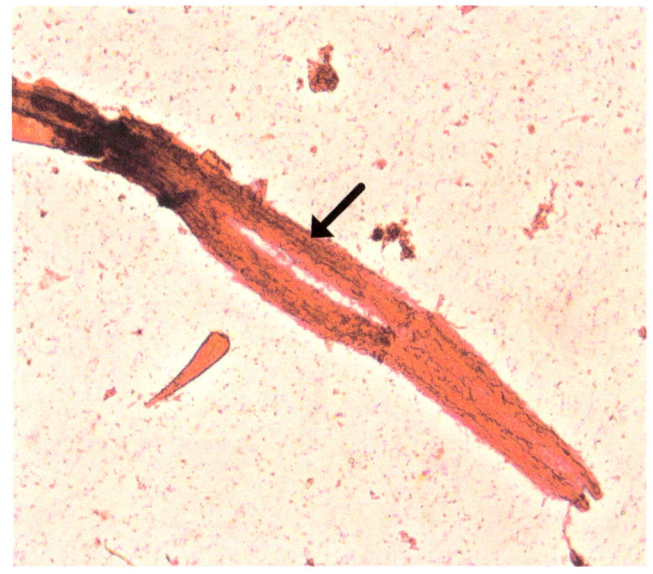

Fig d (above). Crystalline cones. X 200 Magnification. Fig e (right). Stained pickled slide of the ommatidium showing the long tapering tubes, the arrow showing the rhabdom which is just visible. X 100 Magnification.

Fig f. An ommatidium. A. longitudinal section. B. Transversal section. Taken from Dade by kind permission of IBRA.

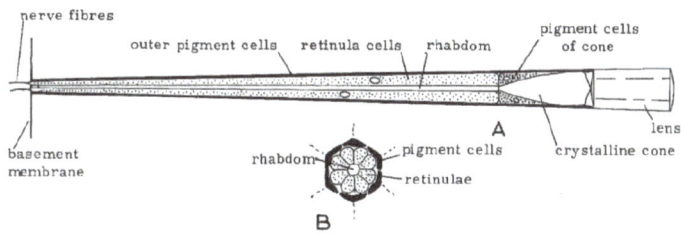

An opportunity to see in ultraviolet!

Scientists have divided the ultraviolet part of the spectrum into three regions - the near-ultraviolet, the far -ultraviolet, and the extreme-ultraviolet; the three regions are distinguished by how energetic the ultraviolet radiation is and by the "wavelength" of the ultraviolet light, which is related to energy. The name comes from the Latin *ultra* meaning 'beyond' and violet being the colour of the shortest wavelengths of visible light at 400nm. UV light has a shorter wavelength than that violet light.

Fig 339. Starting with infrared and finishing with ultraviolet light, the purple arrow shows what colour a bee see and the yellow that of the human's range of vision.

The sun emits ultraviolet radiation in the three bands; however, the Earth's ozone layer blocks 98.7% of this UV radiation from penetrating through the atmosphere.

The UV range of the spectrum has no colour, as perceived by the human eye. Therefore we are free to assign any colour we like, albeit that, under ultraviolet light and depending on the frequency of the lamp used together with any filters in place, the colour ranges from pale violet to almost black. The trick is not so much to compare colours but to compare patterns. Fluorescence may be a common trait to most flowers but might be of temporary occurrence for parts of a flower.

Anthers, style, and pollen grains scattered on the petals are occasionally are seen to fluoresce and strong fluorescence has been noted from the nectar glands. Fluorescence from outside the bracts is exhibited by some species.

Not all flower species, however, have the typical centric UV pattern, which may be confined to symmetrical petal flowers, these might act to attract the bee to the centre of the flower and making them stand out from their surroundings. Some flowers exhibit a virtually endless variety of spectral signatures.

Scientists in the Ecology of Vision group at the School of Biological Sciences (University of Bristol) have argued that much of our understanding of how insects and birds see UV colour is fundamentally flawed.

If one watches a wildlife series with, say, the red light source of your television removed (or if one is red-green "colour-blind) the conclusions one arrives at about colour variation in the natural world, would be unreliable. Yet that is what we are doing every time we think we are seeing the colour world of non-human animals. Unlike other variables such as length, width, mass, or time of day, colour is not an inherent property of the object; it is a property of the nervous system of the animal perceiving the light.

Understanding Colour Vision in Humans

The sensation of colour stems from the differential stimulation of the different types of photoreceptors in the retina. Each cone type produces an output, and it is their differences in output at a particular point on the retina that underlies the sensation of colour.

In humans, there are only three types of cones that absorbing photons in different regions of the spectrum. Due to the appearance of monochromatic light at these wavelengths, these three cone types are called "red", "green" and "blue". Consequently, for humans, all hues can be produced by mixing red, green and blue light. The hue is how we perceive an object's colour, red, orange, green or blue. This is how a colour television set works; a mixture of three wavelengths produces several million apparent "colours". There are several problems, however. Firstly, different wavelength spectra can produce the same hue: as long as the output from the three types of cone remains the same, the hue is the same. Secondly, the same wavelength spectra will produce different hues to animals that differ in the absorption spectra of their cone types. Thirdly, humans have a trichromatic, three-dimensional, colour vision because we have three interacting cone types. Animals with two interacting cone types, such as most mammals other than old-world primates, have a two-dimensional colour vision (similar perhaps to the faulty colour TV set mentioned earlier). It is harder to imagine what colour vision with more dimensions than three might be like, but animals with four and five-dimensional colour visions exist.

When asked to identify the colour of an object, you will most likely speak first of its hue. Quite simply Chroma describes the vividness or dullness of colour; in other words, how close the colour is to either grey or the pure hue. For example, consider the appearance of tomato and radish; the red of the tomato is vivid while the radish appears duller. An interesting factor for me is although I am not colour blind I do have astigmatism in one eye, thus I perceive colour differently in each eye, one being a shade darker than the other. But as they work as a pair, my good eye over-rules the other.

The Colour Vision in Bees.

Bees, like humans, have three receptor types, although unlike humans they are sensitive to ultraviolet light, with loss of sensitivity at the red end of the spectrum. This spectral range is achieved by having one cone type that is sensitive to UV wavelengths and two that are sensitive to 'human visible' wavelengths, blue and green. Remember, because 'colour' is the result of differences in the output of receptor types, bees do not simply see additional 'UV colours'; they perceive even human-visible spectra in different hues to those which humans experience. Fortunately, as any nature film crew knows, we can gain an insight into the bee colour world by converting the blue, red and green channels of a video camera into UV, blue and green channels. Because bees are trichromatic, like us, the three dimensions of bee colour can be mapped onto the three dimensions of human colour.

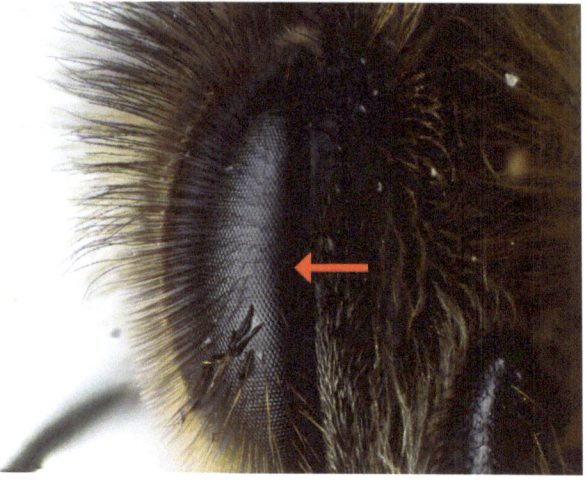

Fig 340. The compound eye of a worker bee, showing the hexagonal facets.

Until relatively recently it was thought that humans had amongst the best colour vision of any animal and that most animals' spectral sensitivities lay within the human-visible spectrum, a misapprehension that persists outside the visual sciences. Other creatures, like butterflies and some reptiles, can see in four or more primary colours.

One of the advantages of UV light for bees and birds is that many fruit, flowers and seeds contrast with their background much more strongly in UV than human-visible wavelengths. As insects evolved in evolutionary time together with plants they have built up many symbiotic methods.

What we see	What bees see	Add in UV
Red	Black	UV Purple
Orange	Yellow-Green	
Yellow	Yellow-Green	UV Purple
Green	Green	
Blue	Blue	UV Violet
Violet	Blue	UV Blue
Purple	Blue	
White	Blue-Green	
Black	Black	

Fig 341. Human and bee comparative colour chart

Thus it has been suggested that flowers developed colour as a way to stand out from the surrounding green foliage as an aid to attracting insects for pollination, with the bonus of an ultraviolet central pattern to indicate the pollen area. Bees are also guided by smell, but that is discussed in another article in the book.

Fig 342-343 (top row). Dandelions (*Taraxacum officinale*) as humans (left) perceive it and as birds (right) might see it. By kind permission of István Bocskai: www.momentslumiere.com
Fig 344-345 (bottom row). Left: how a bee might 'see' when approaching a flower with its compound vision; right: as seen from a distance.

Little eyes

The honeybee has an additional three triangular orientated, simple eyes, known as the dorsal ocelli. Latin for little eye, these are situated on the top of the bee's head, covered by hairs on the worker bee, but larger and more central on the drone due to their larger compound eyes, converging at the top of the head. Much of the bee's behaviour that requires visual input is performed by the compound eyes. If these are covered, the bee cannot fly using the ocelli alone. If the ocelli are covered, they can still fly but start later and finish earlier in the day. The ocelli which are less sensitive to light, differ greatly in structure compared to the compound eyes, being a single biconvex lens made from a thickening of the head cuticle; this, in turn, overlying a retina composed of a layer of about eight hundred photoreceptor cells. The lens does not focus on the retina (its focal point is beyond the retina) so much as acts as a condenser of light. There is a layer of black pigment cells between the lens and retina, which acts as an iris, restricting excessive light on bright days. The cells respond to green and ultraviolet light.

Of the dorsal retinas, the top two view the horizon while the ventral retina, the single bottom one, views the sky, suggesting quite different roles in altitude control. It would appear that the major role of the ocelli is in-flight stability.

Fig 346-347. Drone ocelli with close up image; note the forward position due to the larger compound eyes joining in the centre of the head.

Fig 348. Close up view of an ocellus biconvex lens. X 100 Magnification.

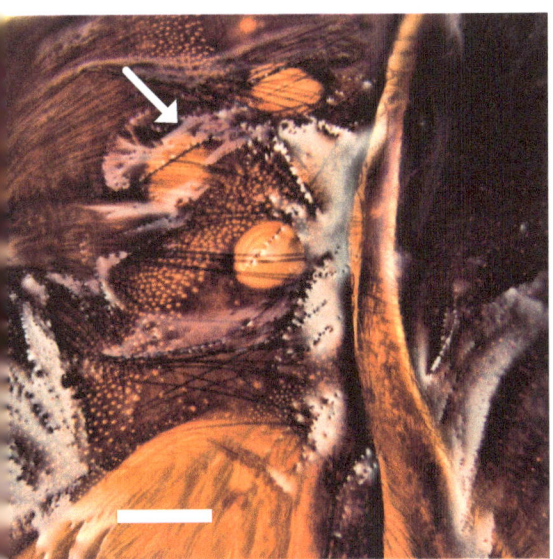

Fig 349. Worker bee head, showing three transparent ocelli, cleared specimen. Scale bar 100 microns.

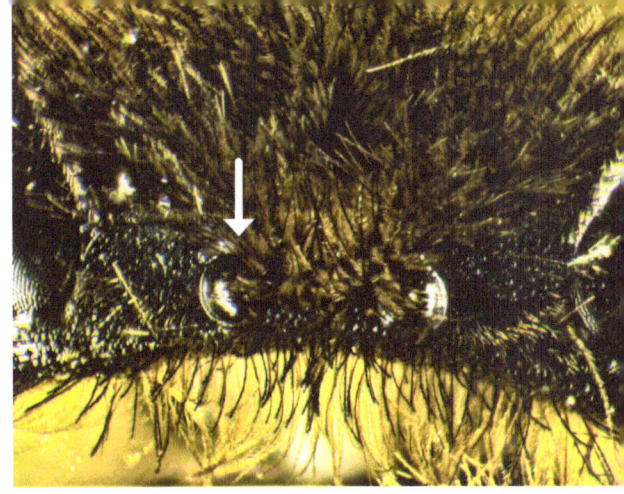

Fig 350. A worker bee, dorsal ocelli, ventral ocelli are hidden by hairs.

Fig 351. A worker bee showing one visible ventral ocellus.

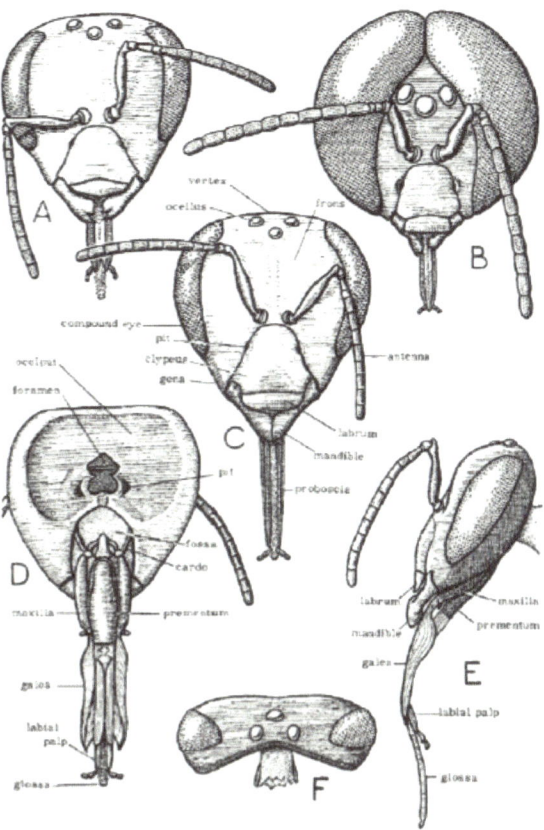

Fig 352. A. Face of the queen. B drone. C. worker. View on the worker's head. D posterior. E. lateral. F dorsal. Note the location of the ocelli. Taken from Dade with kind permission of IBRA.

Pests

Pest (noun.)

1550 "plague, pestilence, epidemic disease," from French *peste* **(the 1530s), from Latin** *pestis* **"deadly contagious disease; a curse, bane," a word of uncertain origin. Meaning "any noxious, destructive, or troublesome person or thing" is attested by c. 1600.**

Uninvited guests

Over the last beekeeping season, I have collected various insects from honey samples that I have been given to try to identify the pollen source and some that have lived in the hive. I think that they are all correctly identified. If we have an insect expert amongst us please step forward!

Fig 353. This was found in a honey sample. With eight legs, this makes it a mite, possibly brought in from a flower visited by a bumblebee, as it seems to be a parasitellus mite of some sort, which are a pest of the bumblebee and can be seen on their body if heavily infested. They do not seem to cause honeybees a problem and are only found in the hive on occasions.

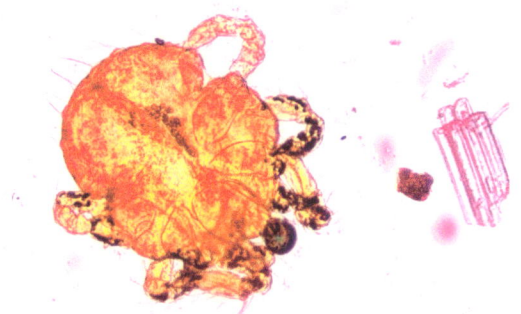

Fig 354. This was found in a honey sample. It is a bee louse, more commonly known as Braula coeca; once a pest, but now rarely seen due to Varroa treatment killing it off. Eggs are laid inside the honey cells before capping; the resulting larvae burrow out and eventually pupate. The adult emerges and likes to find the queen being a permanent member of the colony. They feed on the queen's saliva and are not harmful unless they become excessive, decreasing the efficiency of the queen.

Fig 355 (left). These are pollen mites, taken from the bottom board; not normally a problem for honeybees. They will eat pollen stores.

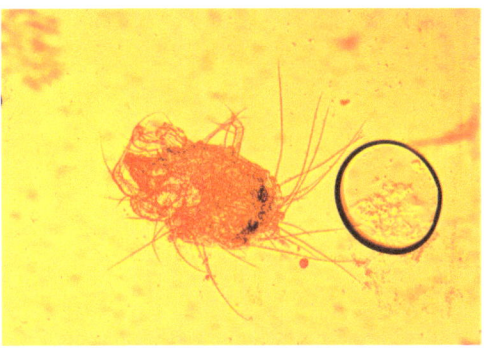

Fig 356. Squashed Acarine mite.

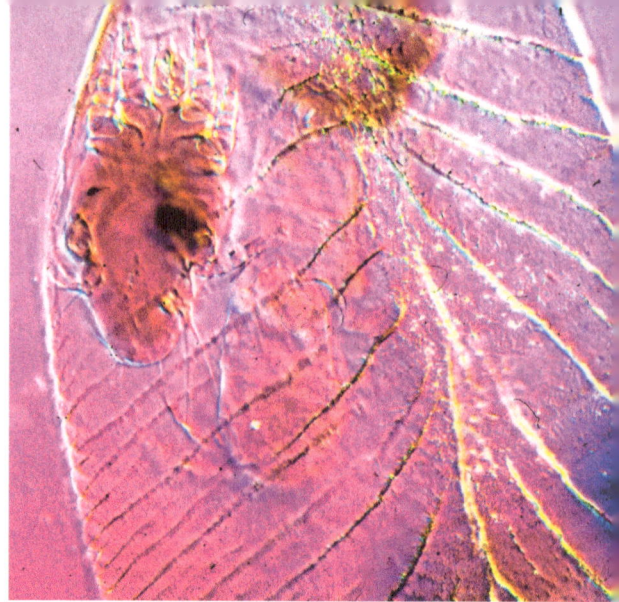

Fig 357. Acrine mites inside of trachea. © Dr Michel Asperges. Universiteit Hasselt Agoralaan gebouw D te BE 3590 Diepenbeek, Belgium

Acarine mite taken from a honey sample, I have never seen this before. I think that the Varroa treatment is controlling these insects; the hive it was found in had been treated regularly for Varroa.

Once the scourge of beekeepers gains entry to the trachea via the first spiracle, where it bites & feeds from the bee's haemolymph. Blamed for the Isle of White disease in 1920, although this is now in doubt.

The Roman statesman and scholar, Marcus Terentus Varro, who served as Julius Caesar's librarian, drafted a theory known as the 'Honeycomb Conjecture'. Being a beekeeper he noted the hexagonal shape of the comb and proposed they built them that way for the sake of efficiency. No other inter-locking shape would hold so much honey with so little wax. It was not until 1999 that a mathematician proved him right. Taxonomists came up with the genus 'Varroa' when naming the dreaded mite, forever associating the old Roman with the deadly threat to the bees he so admired.

Over 90 differing mites infest honeybees in Europe - most do no harm.

Fig 358. Developing Varroa mites on pupa. © Dr Michel Asperges. Universiteit Hasselt Agoralaan gebouw D te BE 3590 Diepenbeek, Belgium

Fig 359. Taken from the bottom board, the dreaded female Varroa mite, a member of the spider family. Public enemy number one.

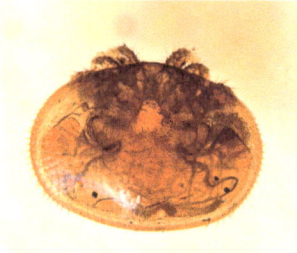

Fig 360. Varroa without its Toga! X 30 Magnification.

Fig 361. Male Varroa mite.

Fig 362. Varroa mites in the bottom of the cell, note brown female in the background and developing one in the foreground coloured white.
© Dr Michel Asperges. Universiteit Hasselt Agoralaan gebouw D BE 3590 Diepenbeek, Belgium.

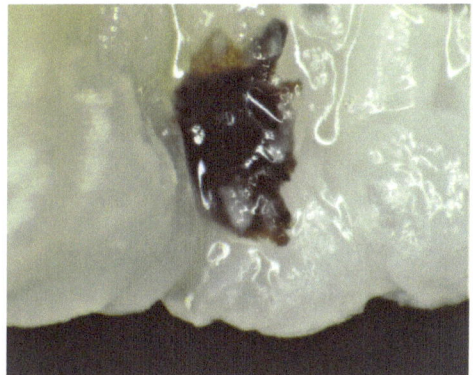

Fig 363. Developing Varroa mite buried inside of a drone larva.

Let us hear it for the Wax Moth

The wax moth gets bad press as far as the beekeeper is concerned. It lays its larva in the wax comb from which it proceeds to devour the wax, leaving a vast number of tunnels as evidence; this can lead to the death of the colony if the infestation is a heavy one.

The Greater Wax moth, Galleria mellonella, is the more destructive and common pest whilst the Lesser Wax moth, Achroia grisella, is both less prevalent and less destructive. As usual in nature there always seems to be a balance between host and pest, otherwise - the host would soon die out and take the pest with it.

Not all relationships are symbiotic or even instantly recognisable, nor do they appear good for the host, but the wax moths do clear up old colony sites left by swarming bees in the wild, removing all that remains including not only the wax but also the silk and waste that the honeybee larva had deposited over time in the comb. Think of them as super cleaners!

Scientists have used these critters as an aid in medical research for the greater good of humankind. Wax moth larva change colour when they are sick, an indicator which makes them valuable in testing for bacterial concentrations related to human. A problem can be indicated by a ratio of 1,000 bacteria per millilitre or 10 million bacteria from which the infection rate can be defined.

As a practical example, it is recommended to wash fruit before we consume them; this does not remove all the bacteria but dilutes them to a more manageable level for our bodies.

**Fig 364. Newly emerged Wax moth.
By kind permission of Richard Ball.**

Fig 365. The resulting damage done by the wax moth silk, black specks are frass, droppings. By kind permission of Richard Ball.

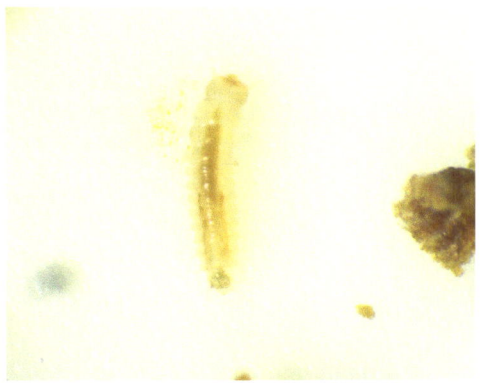

Fig 366. This is a Wax moth larva taken from the bottom board and will cause major problems and destroy a hive if infested in moderate numbers. There are two types found in the UK, the lesser and greater wax moth.

Fig 367. Fully grown larva.

Could wax moth larvae be the next superfood? We in the west do not seem to eat insects yet, but with the world exponential growth of people, we may well have to!

The common wasp, Vespula vulgaris

The much-maligned wasp is an important insect and normally only bothers the colony in the autumn when their larvae are all gone and they are seeking a sweet reward whereever they can find one. This need is normally fulfilled when the returning wasp's exchanges the insects caught, for a reward from the larvae, a sweet sugary secretion. Once all of the larvae have matured this is no longer available, but the craving for it lingers on.

**Fig 368. Wasps can be identified by their facial features and abdominal patterns.
Fig 369. Wasp resting in pear tree near hives.
Fig 370. Wasp using its arolium pads to grip the glass.
Fig 371. Wasp hawking by hive entrance.**

Glossary

Abdomen	An area containing the digestive and reproductive organs.
Albuminoidal	Type of protein.
Alleles	Alternative forms of a gene that arises by mutation and is found at the same place on a chromosome.
Antennae	Latin antemna, horizontal mast spar, the main sensory organ of the bee.
Anus	The external opening of the intestine. Ring or circle.
Aorta	A strap to hang something by, the main artery of the heart.
Apodeme	The frame of the body.
Arolium	Lobe. Pad on the pretarsus.
Axon	Axis, nerve fibre.
Bastitarsus	Base of tarsus.
Bulb of endophallus	Penis bulb.
Bursa copulatrix	Mating pouch.
Cardines	A hinge. Extends proboscis.
Cervix of endophallus	Neck of endophallus. Chitinized plates. Hardened plates. Tunic or shield.
Chorion	Membrane.
Chromosome	Found in every cell nucleus, divide into new sets.
Cibarium	Food chamber.

Cornua	Horns found on the endophallus, coloured orange on mature drones.
Corpora allata	Body moving forward to the final position.
Corpora cardiaca	Bodies near the heart.
Copula	Mating.
Cubital	Elbow.
Cuticular intima	Cuticle coating. Cytoplasm. The fluid inside a cell.
Diaphragm	Partition, barrier, membrane.
Diploid	Di means two, and ploid stands for chromosome, having two sets.
Discoidal	Flat disc shape.
Ecdyses	Greek, to strip or moult.
Ejaculatory duct	The tube leading from the seminal vessels to the bulb conveys the sperm and mucus. Endocrine. In to the body.
Endophallus	Meaning a 'penis held within'.
Entomology	Insect.
Enzyme Leavened	A catalyst.
Epithelium	A thin layer of skin.
Eversion	The pushing out and turning inside out of the endophallus.
Exoskeleton	Protective, hardened exterior.
Filiform	Threadlike.
Fissure	A groove. Fimbriate lobe. A petal-like structure, acting as a balloon when turned inside out.

Flabellum	A little fan.
Flagellum	Latin for a little whip, the end section of the antennae.
Fossa	A trench.
Furca	Pronged fork.
Fusiform	Spindle.
Galeae	Greek for a helmet.
Ganglion cells	A tumour or swelling.
Gaster	A term for the abdomen, taken from the Greek for the belly.
Gene	Unit of inheritance found in every chromosome.
Geniculate	Knee.
Genotype	The genetics of a bee or colony a set of inherited genetic instructions encoded in its DNA.
Glossa	Tongue.
Glycogen	Sugar producer.
Gonopore	Opening of the ejaculatory duct.
Haploid	Greek word haplos meaning single, containing only one set of chromosomes.
Haemolymph	Blood of the bee.
Hemizygote	A diploid individual with only one allele at a given locus or gene instead of the typical two.
Homozygous	Processing only an identical set of chromosomes as the mother.
Hypodermis	Below the skin.

Hypopharyngeal	Under the pharyngeal, throat.
Imago	Adult honeybee, the final stage of growth.
Instar	The phase between two stages of growth.
Labium	Lower lip.
Labrum	Upper lip.
Laciniae	A flap or fringe.
Lamina	Thin plates.
Lamina parameralis	Upper claspers, used in other insects when mating, too small to be of any use in bees.
Larva	Disguise. Immature stage of honeybee.
Lateral oviduct	Top duct connecting to the ovaries.
Lipid	Fats.
Lorum	A thong.
Lumen	Cavity.
Malpighian tubules	Excretory organs of the bee.
Mandibles	Upper jaw.
Mating sign	Part of the endophallus left behind in the queen's sting chamber.
Maxilla	Lower jaw.
Median oviduct	Duct where eggs pass through.
Medulla	Marrow.
Mesosoma	Middle body.
Metasoma	After body or mesenteron.
Micropyle	Small gate.

Morphs	A genetic variant of an animal.
Mucous glands	Producers of mucus.
Mucus	Part of the final ejaculate. Neotenin. Juvenile hormone.
Nerve	Sinew.
Ocelli	Simple eye.
Ommatidium	A little eye.
Oenocytes	Secretory cells.
The organ of Johnson	Found in the antennae used for determining airspeed when in flight and sound detection.
Ostia	A door. One way valves of the heart.
Ovarioles	A little ovary. Consists of the egg, nurse and follicle cells.
Ovary	An egg. Consists of about 150 ovarioles.
Paraglossae	By the side of the tongue.
Parthenogenesis	Sexual reproduction without fertilisation.
Pedicel	A small stalk-like structure, part of the antennae.
Peritrophic	A tubular chitinous sheath that lines the midgut.
Petiole	A stalk. The waste.
Phallotreme	Opening for the endophallus and anus.
Phragma	A fance. For muscle attachment.
Pharynx	Throat. The gullet.
Phenotype	The set of observable characteristics.
Pheromone	Chemical messages that are produced by the bee to communicate.

Planta	Sole of foot.
Pleurites	A rib.
Pneumophyses of endophallus	Lateral lobes.
Polarised light	A light that is vibrating in more than one plane, not visible to humans.
Post-eclosion	After hatching from the cell.
Postmentum	Behind the chin.
Pretarsus	The foot of an insect.
Primordial germ cells	The biological cell that gives rise to the gametes of an organism that reproduces sexually.
Proboscis	Feeder. A collection of mouth parts.
Proctodaem	Hindgut.
Proventriculus	Before the belly.
Pupae	Stage in the development when complete metamorphosis between larvae and egg. Ramus. A branch or twig.
Ratellum	A little rack.
Retina	Net like tunic.
Retinotopy	A place.
Retinula	Little retina.
Rhabdom	A rod.
Rectum	The straight ending of the intestine.
Sac of the bulb	Orifice. Scape. Rigid stalk, connecting with the head joint.

Sclerites	The outer covering of the bee, its exoskeleton made up of chitin.
Scutellum	Little shield.
Semen	Mixed ejaculatory fluid. Seminal vesicle. Sperm storage vessel.
Sensilla picola	Specialised individual sense organs.
Sperm	Seed.
Spermatheca	Queen's sperm storage organ.
Spermatogenesis	The process of sperm development.
Spiracle	To breathe.
Sternite	Plates covering the abdomen.
Sternum	Firm or solid.
Stomodaeum	Foregut.
Stipites	A stack or trunk.
Suboeaophageal	Under oesophagus.
Sulci	Furrow.
Symbiotic relationship	The mutual benefit of both parties.
Taenidia	Ribbon.
Tarsal	Ankle or instep.
Tarsomere	Apart from the body.
Tentorial	Tent pole.
Tergum	Back.
Tergite	Part of the back.

Testes	Latin for witness. The organ where sperm are made.
Thorax	Torso.
Trachea	The windpipe.
Tracheoles	Windpipe, smaller.
Triglycerides	Type of fats.
Trophocytes	A fat cell that supplies nourishment.
Tubules	Small pipe.
Ungues	Claws.
Urate	Salt of uric acid. Waste from ammonium acid.
Vasa deferentia	Vessels that convey sperm.
Venom	Poison.
Ventriculus	The midgut and true stomach where digestion takes place.
Vestibulum at base of the endophallus	A large internal cavity.

References

Textbooks used for references throughout this book

Anatomy and Dissection of the Honeybee. Dade. IBRA. 2009. ISBN 0-86098-214-9

Anatomy of the Honeybee. Snodgrass. Cornell University Press .1984 IBSN 978-0-8014-9302-7

Beekeeping study notes. J D Yates and BD Yates. 1996. BBNO ISBN 0 905652 33 9

Beyond Extreme Close-Up Photography. Julian Cremona. 2018. The Crowood Press Ltd ISBN 978 1 78500 645 0

Form and function of the honeybee. Lesley Goodman. IBRA. 2003 ISBN 0 86098 243 2

Insect Microscopy. Andrew Chick. The Crowood Press. 2016. ISBN 987 1 78500 201 4

Morphology of the Honeybee Larva. Nelson. 1924. Available on the internet for free. National Agricultural Library Digital Collections. Journal of agricultural research p. 1167-1214. 1924. Rights. This item is authored by federal employees as part of their official duties and are therefore non-copyrightable and/or published by the federal government and now in the public domain.

Practical Microscopy. Eric Marson. Northern Biological Supplies. 1983.

Practical Microscopy for Beekeepers. Bob Maurer. BeeCraft. 2012. ISBN 9 780900 147135

Queen Bee: Biology, Rearing and Breeding. David Woodward. Northern Bee Books. 2010. ISBN 9 781904 846352

Safe Microscopic Techniques for Amateurs. Slide Mounting. Walter Dioni Onview.net Ltd 2014 ISBN 9 781499 746518

The Buzz about Bees. Tautz. Springer. 2008. ISBN 978-3-540-78727-3.

The Embryology of the Honey. J A Nelson. 1915 Princeton University Press. Available on the internet for free.

Understanding Bee Anatomy, a full-colour guide. Stell. The Catford Press. 2012. ISBN 978-0-9574228-0-3

Websites quoted

Dave Cushman's web site - please visit to find more excellent information on any aspect of beekeeping. www.dave-cushman.net. By kind permission of Roger Patterson in whose safe hands Dave left the site for future beekeepers.

Online Etymology Dictionary | Origin, history and meaning of English words (etymonline.com)

Papers quoted

Larva and Pupae

Conservation of Essential Design Features in Coiled-Coil Silks

Tara D. Sutherland, Sarah Weisman, Holly E. Trueman, Alagacone Sriskantha, John W. H. Trueman, Victoria S. Haritos

Molecular Biology and Evolution, Volume 24, Issue 11, November 2007, Pages 2424-2432, https://doi.org/10.1093/molbev/msm171

Published: 16 August 2007

Open Access article distributed under the terms of the Creative Commons Attribution Non-Commercial License (http://creativecommons.org/licenses/by-nc/2.0/uk/) which permits unrestricted non-commercial use, distribution, and reproduction in any medium, provided the original work is properly cited.

Hierarchical, multilayered cell walls reinforced by recycled silk cocoons enhance the structural integrity of honeybee combs Kai Zhanga, B Huiling Duana, Bhushan L. Karihaloob, 1, and Jianxiang Wanga State Key Laboratory for Turbulence and Complex Systems and Department of Mechanics and Aerospace Engineering, College of Engineering, Peking University, Beijing 100871, China; and b School of Engineering, Cardiff University, Queen's Buildings, The Parade, Cardiff CF24 3AA, United Kingdom

Rozen, Jerome G Jr et al. "Survey of Hatching Spines of Bee Larvae Including Those of Apis mellifera (Hymenoptera: Apoidea)." Journal of insect science (Online) vol. 17, 4 (2017): 89. doi:10.1093/jisesa/iex060

This is an Open Access article distributed under the terms of the Creative Commons Attribution License (http://creativecommons.org/licenses/by/4.0/), which permits unrestricted reuse, distribution, and reproduction in any medium provided the original work is properly cited.

H. Hepburn, Orawan Duangphakdee, Christian Pirk. Physical properties of honeybee silk: a review. Apidologie, Springer Verlag, 2013, 44 (5), pp.600-610. ff10.1007/s13592-013-0209-6ff. ffhal-01201329f

Heart of the matter

The antennal circulatory organ of Apis mellifera and other Hymenoptera: function morphology and phylogenetic aspects. S Matus G.Pass International Journal of Morphology and Embryology. January 1999.

Comparing micro-CT results of insects with classical anatomical studies: The European Honeybee (Apis mellifera Linnaeus, 1758) as a benchmark (Insecta: Hymenoptera, Apidae) ©Prof. Dr Javier Alba-Tercedor. Univ. of Granada. Spain https://analyticalscience.wiley.com/do/10.1002/micro.2789/full/

Who was Malpighi?

Open access insect physiology. www.dovepress.com/comparative-physiology-of-malpighian-tubules-form-and-function-peer-reviewed-fulltext-article-OAIP

BeeBase UK   www.nationalbeeunit.com

Mandibles are made for many things

Proboscis, a drinking straw by another name

Drawing of the formed proboscis showing section at the top, the channels. By kind permission of D G Mackean www.biology-resources.com

Proventriculus: before the belly

The Action of the Proventriculus of the Worker Honeybee, Apis Mellifera.

By L. Bailey. Bee Research Department, Rothamsted Experimental Station

Honeybee drones possessors of some very specialised equipment

Source: Prof J Woyke by kind permission.

Designed for all events

The pretarsus of the honeybee doi.org/10.26496bjz.2017.8)

Acknowledgements

A debt of gratitude must go to Professors Woyke, Tautz and Seeley for kindly allowing me to use their articles and research data.

IBRA has the copyright for the following books and articles. Dade's Anatomy & Dissection of the Honeybee. Bee World. Form and function of the honeybee. Lesley Goodman. Permission to use the data and images has been kindly granted and acknowledged.

Understanding Bee Anatomy, a full-colour guide. By Doctor Ian Stell.

My external thanks go to all the bee scientists who have freely published their research. The many beekeepers who have educated me in a multitude of ways, especially Master Beekeeper Chris Utting, Julie Elkin, Lilah Killock, Glyn Davis, Richard Simpson who keeps me on my beekeeping toes and Alan White for all those bee thoughts over a beer or two. Jeremy Barnes and Colin Wood for reading the text and making many helpful suggestions.

My wife Catherine Kingham the queen in my life.

I am in constant debt to Pat and David Woodward who are both highly qualified microscopists. They have helped and advised me over the years and have always encouraged me to pursue my interests, once again, a very big thank you.

Norman Carreck BSc CBiol FRSB FRES NDB for all the support and leads over my beekeeping career, no matter what the questions, plus encouragement, papers, solutions and contacts that have been freely given, without which none of my books would have been possible.

A special thank you must go to the following Professors:
Professor Dr Michel Asperges. Universiteit Hasselt Agoralaan gebouw D te BE 3590 Diepenbeek, Belgium for allowing me to use his superb images and for all the pollen information.
Professor Dr Klaus Hartfelder Faculdade de Medicina de Ribeirão Preto Brazil for his extensive help throughout this project. I am grateful for your time and effort in making this a better publication.

All remaining shortcomings are my own responsibility.

All photographs by Graham Kingham unless otherwise noted.

Technical

For those of a technical mind the microscopes used were: Meiji compound biological microscope ML4200H

Brunel zoom dissecting trinocular stereomicroscope BMDZ. Brunel has a vast range of quality microscopes new and used suitable for every pocket. They offer outstanding service and advice on all matters concerning the microscope world. They also supply chemicals, stains and a lot more including microscope cameras. www.brunelmicroscopes.co.uk

Camera used: Toupcam E3CMOS. 12 megapixels. USB 3. Software: ToupView.

Olympus OM D E-M1 camera with a 60mm macro lens.

Lighting

Lighting is reflected using LED with diffusers on. Transmitted lighting is LED or tungsten, Tungsten lighting was also used at a low light level with a blue filter to reflect daylight conditions. Polarising filters have been used on some specimens.

Dissecting

Most insect books give details to clear the soft interior organs for preservation; there are several means of doing this, see reference books for details. The paper quoted below offers another option, which is quicker and gives good results using chemicals available to the amateur microscopist.

Clearing and dissecting insects for internal skeletal morphological research with particular reference to bees.

Diego Sasso Portoa, Gabriel A.R. Melob, Eduardo A.B. Almeidaa

© 2015 Sociedade Brasileira de Entomologia. Published by Elsevier Editora Ltda. This is an open-access article under the CC BY-NC-ND license (http://creativecommons.org/licenses/by-nc-nd/4.0/).

Some bees have been pickled in alcohol to preserve them and to help show different bodily parts, which are not always as clear in a fresh bee.

For photographic reasons, mainly light diffraction and reflection, most specimens have been dissected dry and not fixed in wax and supported in a water dissecting fluid, I have used dense cell, white foam to pin the bee in place. As usual, practice, patient and an opportunist moment when something goes wrong, often produces a good result, even if it is not the thing you set out to achieve.

The methods of specimen preparation

The specimens have been photographed when alive. Most of the dissected specimens have been collected outside the hive, early evening, from a tray at the bottom of the entrance, where they have died. Preserved in the fridge for up to 3 days.

I have dissected these dry rather than in water due to the reflection from the lights causing diffraction and white spots.

Some bees have been preserved in alcohol; others have been cleared of internal organs, and some frozen.

Slides have also been prepared by traditional methods. See the book list for further details.

Index

Abdomen	77
Albuminoidal	130
Antennae	63 127 144
Anus	28 83
Aorta	36
Apodeme	79
Arolium	125
Axon	60
Bastitarsus	125
Bulb of endophallus	100
Bursa copulatrix	180
Cardines	80 91
Cervix of endophallus	99
Chitinized plates	86
Chorion	14 108
Chromosome	104
Cibarium	92 134
Cornua	101
Corpora allata	23
Corpora cardiaca	23
Cubital	114
Cuticular intima	16 54 146
Cytoplasm	14
Diaphragm	38
Discoidal	114
Ecdyses	
Endocrine	21 68
Endophallus	42 99
Entomology	82
Enzyme	134
Eversion	99
Exoskeleton	56 66 77
Filiform	92
Fissure	71 120
Flabellum	98
Flagellum	146
Fossa	90
Furca	78
Fusiform	16
Galeae	93
Ganglion	29 60
Gaster	73 77
Gene	103
Geniculate	144
Genotype	113
Glossa	91
Glycogen	130
Haemolymph	66 82
Hypodermis	34
Hypopharyngeal	134
Imago	16 38
Instar	38 45
Labium	18
Labrum	18 134
Laciniae	90
Lamina	152
Lamina parameralis	152
Larva	16
Lipid	22 97
Lorum	90
Lumen	31 132
Malpighian tubules	28 31 82
Mandibles	88
Mating sign	99
Maxilla	18
Mesosoma	73

Metasoma	73
Micropyle	108
Morphs	123
Mucous glands	103
Mucus	102
Nerve	29
Ocelli	158
Ommatidium	152
Oenocytes	26 130
The organ of Johnson	150
Ostia	37 66
Ovarioles	107
Ovary	38 107
Paraglossae	90
Pedicel	146
Peritrophic	95
Petiole	67 73
Phallotreme	99
Phragma	120
Pharynx	28
Phenotype	113
Pheromone	123 132 136
Planta	125
Pleurites	77
Pneumophyses of endophallus	104
Post-eclosion	42
Postmentum	73 90
Pretarsus	123
Primordial germ cells	41
Proboscis	90
Proctodaem	28
Proventriculus	91
Pupae	22
Ramus	141
Ratellum	128
Retina	153
Rectum	85
Scape	146
Sclerites	121
Scutellum	71
Semen	101
Seminal vesicle	43 102
Sensilla	123
Sperm	102 145
Spermatheca	104
Spermatogenesis	41
Spiracle	34 53
Sternite	77
Sternum	53
Stomodaeum	14 28
Stipites	90
Suboeaophageal	63
Sulci	77
Taenidia	57
Tarsal	124
Tarsomere	124
Tentorial	79
Tergum	53
Tergite	77
Testes	38
Thorax	71 120
Trachea	33 56
Tracheoles	33 56
Triglycerides	130
Trophocytes	26 130
Tubules	22 31 82
Ungues	123
Urate	26
Vasa deferentia	38
Venom	61 141
Ventriculus	94
Vestibulum at base of the endophallus	101

Other books by Graham Kingham

The Honeybee Drones: Specialists in the Field

This book concentrates only on the drone – the male honeybee.

It provides details regarding the drone's internal and external anatomy, production and development, behaviour, role in the hive, genetics and more. Copiously illustrated, the book also discusses the latest research updates on drones.

Published by Northern Bee Books Ltd.

Visit www.northernbeebooks.co.uk for the world's largest collection of English language bee books, both new and second-hand.

The Honeybee: A Hive of Information.
Annotated Anthology

- Do bees sleep?
- What is in bee poo?
- Do they have knees?
- What about the birds and the bees'?
- Why do bees have a hump?

Many of the honeybee's life mysteries are answered here. Micrographs of pests in your honey and diseases in the hive. Some contentious thoughts on the future management of the Apis mellifera species. Short readable articles to dip in and out of for reference and amusement.